Modernism and ...

Series Editor: Roger Griffin, ... kes University, UK.

The series *Modernism and...* inv... phenomena t ... in mo... and 'modernism'. Apart from ... but ground... ecialist monographs, the books aim through their cumulative ... the application of this highly contested term beyond its conven- ... and aesthetics. Our definition of modernism embraces the vast ... ve acts, reforming initiatives, and utopian projects that, since the ... tury, have sought either to articulate, and so symbolically tran- ... malaise or decadence of modernity, or to find a radical solution to ... ent of spiritual, social, political – even racial – regeneration and ... nate aim is to foster a spirit of transdisciplinary collaboration in ... tural forces that define modern history beyond their conventional ... eworks.

...ND EUGENICS

...ND NIHILISM

... AND THE EUROPEAN NEW RIGHT

...NISM AND GENDER

onnelly
...ISM AND THE GROTESQUE

...arling
...ISM AND DOMESTICITY

...ldman
...SM AND PROPAGANDA

...M AND MEDITERRANEANISM

Griffin
...ERNISM AND TERRORISM

Ben Hutchinson
MODERNISM AND STYLE

Carmen Kuhling
MODERNISM AND NEW RELIGIONS

Patricia Leighten
MODERNISM AND ANARCHISM

Thomas Linehan
MODERNISM AND BRITISH SOCIALISM

Gregory Maertz
MODERNISM AND NAZI PAINTING

Paul March-Russell
MODERNISM AND SCIENCE FICTION

Anna Katharina Schaffner
MODERNISM AND PERVERSION

Richard Shorten
MODERNISM AND TOTALITARIANISM

Mihai Spariosu
MODERNISM, EXILE AND UTOPIA

Roy Starrs
MODERNISM AND JAPANESE CULTURE

Erik Tonning
MODERNISM AND CHRISTIANITY

Veronica West-Harling
MODERNISM AND THE QUEST

Modernism and...
Series Standing Order ISBN 978–0–230–20332–7 (Hardback)
978–0–230–20334–4 (Paperback)
(outside North America only)

You can receive future titles in this series as they are published by placing a standing order. Please contact your bookseller or, in case of difficulty, write to us at the address below with your name and address, the title of the series and the ISBN quoted above.

Customer Services Department, Macmillan Distribution Ltd, Houndmills, Basingstoke, Hampshire RG21 6XS, England

MODERNISM AND NIHILISM

Shane Weller

© Shane Weller 2011

All rights reserved. No reproduction, copy or transmission of this publication may be made without written permission.

No portion of this publication may be reproduced, copied or transmitted save with written permission or in accordance with the provisions of the Copyright, Designs and Patents Act 1988, or under the terms of any licence permitting limited copying issued by the Copyright Licensing Agency, Saffron House, 6–10 Kirby Street, London EC1N 8TS.

Any person who does any unauthorized act in relation to this publication may be liable to criminal prosecution and civil claims for damages.

The author has asserted his right to be identified as the author of this work in accordance with the Copyright, Designs and Patents Act 1988.

First published 2011 by
PALGRAVE MACMILLAN

Palgrave Macmillan in the UK is an imprint of Macmillan Publishers Limited, registered in England, company number 785998, of Houndmills, Basingstoke, Hampshire RG21 6XS.

Palgrave Macmillan in the US is a division of St Martin's Press LLC, 175 Fifth Avenue, New York, NY 10010.

Palgrave Macmillan is the global academic imprint of the above companies and has companies and representatives throughout the world.

Palgrave® and Macmillan® are registered trademarks in the United States, the United Kingdom, Europe and other countries.

ISBN 978–0–230–23103–0 hardback
ISBN 978–0–230–23104–7 paperback

This book is printed on paper suitable for recycling and made from fully managed and sustained forest sources. Logging, pulping and manufacturing processes are expected to conform to the environmental regulations of the country of origin.

A catalogue record for this book is available from the British Library.

Library of Congress Cataloging-in-Publication Data
Weller, Shane.
Modernism and nihilism / Shane Weller.
   p. cm.
ISBN 978–0–230–23104–7   (pbk.)
1. Philosophy, Modern. 2. Nihilism. I. Title.
B791.W45 2011
149'.8—dc22   2010034765

10 9 8 7 6 5 4 3 2 1
20 19 18 17 16 15 14 13 12 11

Printed and bound in Great Britain by
CPI Antony Rowe, Chippenham and Eastbourne

In memory of my father
John Jeremy Weller (1938–2010)

Glasgow City Council Cultural & Leisure Services Libraries Info. & Learning **L**	
C 004630358	
Askews	22-Dec-2010
149.8 LL	£16.99

CONTENTS

Introduction: Modernity, Modernism, Nihilism	1
Part I: Philosophical Modernism and Nihilism	**15**
1. From the French Revolution to Nietzsche	17
2. Nietzsche's Long Shadow	42
Part II: Aesthetic Modernism and Nihilism	**75**
3. From Flaubert to Dada	77
4. Kafka and After	102
Part III: Postmodernism and Nihilism	**137**
5. Our Only Chance?	139
Bibliography	166
Index	177

INTRODUCTION: MODERNITY, MODERNISM, NIHILISM

In the field of modernism studies, it has long been the accepted practice to consider modernism primarily, if not exclusively, as an aesthetic phenomenon. As Roger Griffin argues in *Modernism and Fascism* (2007), however, there are also philosophical and political forms of modernism, and if one wishes to move towards a more comprehensive understanding of the movement, then one has to analyse all three forms in a manner that takes account of their relation to one another. The second major claim made by Griffin in his work on fascism is that modernism in this broader sense is to be understood as a reaction against a modernity that is seen to have passed from a revolutionary, progressive phase in the late eighteenth and first half of the nineteenth century, to a decadent and ultimately nihilist phase in the later nineteenth and first half of the twentieth century. In short, modernism is a 'revolt against decadence', an attempt both to destroy that which, in the realms of philosophy, politics, and aesthetics, no longer effectively bestows shape and meaning on experience, and to find 'new sources of meaning, spirituality, and communality' (Griffin 2007: 52). At their most fundamental level, all forms of modernism, be they philosophical, political, or aesthetic, are committed to the idea of palingenesis, to the rebirth of culture in a form that is uncontaminated by the spiritual sickness besetting modernity.

As David Harvey remarks, however, modernism is 'a troubled and fluctuating aesthetic response to conditions of modernity produced by a particular process of modernization' (Harvey 1989: 98). These fluctuations are nowhere more evident than in the history of modernism's relation to the concept of nihilism. If one is to begin to grasp in its complexity the relationship between modernity, modernism, and nihilism as a whole, then, as we shall see, the conjunction of modernism *and* nihilism has to be understood in two, seemingly antithetical ways: on the one hand, modernism *versus* nihilism, and, on the other hand, modernism *as* nihilism. Before one can begin to analyse the various, historically specific ways in which modernism and nihilism are thought as being opposed to, or identical with, one another, however, it is of course first necessary to establish preliminary definitions of our three key terms.

Modernity

Although the debate concerning both the nature and the historical parameters of modernity is far from having been settled, for the purposes of the present work modernity is to be understood as that epoch in which the dominant values are those of the eighteenth-century Enlightenment, with the core value being reason. Griffin offers the following helpful list of the principal elements constituting modernity, including the crucial role played by technology therein:

> the spread of rationalism, liberalism, secularization, individualism, and capitalism, the cult of progress, expanding literacy rates and social mobility, urbanization and industrialization, the emergence of the urban middle class (capitalist) and the working (rural and proletarian) classes from a feudal structure of society, the growth of representative government and bureaucratization, revolutionary developments in communications and transport, geographical discoveries and imperial expansion, the advance of secular science and ever more powerful technology and technocracy. (Griffin 2007: 45–6)

According to Fredric Jameson, modernity thus conceived may be described as a 'catastrophe', since it 'dashes traditional structures and lifeways to pieces, sweeps away the sacred, undermines immemorial habits and inherited languages, and leaves the world as a set of raw materials to be reconstructed rationally' (Jameson 1994: 84). This conception of modernity derives in large part from that of the Frankfurt School, and, in particular, from Max Horkheimer and Theodor Adorno's *Dialectic of Enlightenment* (1947). As Jameson observes of this key work of philosophical modernism, in which modernity is submitted to the harshest of dialectical critiques: 'the scientific ethos of the *philosophes* is dramatized as a misguided will to power and domination over nature, and their desacralizing program as the first stage in the development of a sheerly instrumentalizing world-view which will lead straight to Auschwitz' (Jameson 1998: 25). According to Horkheimer and Adorno, modernity finds its consummation in Nazism as a form of radical nihilism the origins of which lie in the eighteenth-century Enlightenment, which, they argue, embraces both Kant and Sade. While modernity may have undertaken 'the disenchantment of the world; the dissolution of myths and the substitution of knowledge for fancy', it has also led inexorably to a world that 'radiates disaster triumphant' (Horkheimer and Adorno 2002: 3). But if, on the one hand, the Holocaust can be seen as the consummation of the nihilism of *modernity*, it can, on the other hand, also be seen as the consummation of the nihilism of *modernism* as a reaction against the very conditions of modernity. This latter view is the one taken by Griffin, who sees the Holocaust as an extreme consequence of the logic of 'creative destruction' to which fascism (as a form of political modernism) committed itself (Griffin 2007: 182).

As for the commencement of the epoch of modernity, most commentators date this to the late eighteenth century in Europe and to what Griffin terms the emergence of 'the reflexive mode of historical consciousness which legitimated the French revolutionaries' fundamentalist war against tradition' (51). There are, however, those – among them some of the key deployers of the concept of nihilism in their modernist critique of modernity – who locate the origins of modernity much further back. Indeed, in *The Birth of Tragedy* (1872), Friedrich Nietzsche, who is undoubtedly the most important figure in the history of the deployment of the concept of nihilism, divides the history of the West into two principal epochs: the 'tragic' age of pre-Socratic Greek culture, and the 'modern' age that commenced with Socrates as the first embodiment of 'theoretical man', committed to reason as the sole means to achieve an understanding of the world. Martin Heidegger, whose thinking of the nihilism of modernity is profoundly indebted to Nietzsche, makes a similar distinction between two epochs: pre-Socratic Greek and modern.

If the origins of modernity are a matter of dispute, the same is also true of its end. The most widely accepted view is that a transition from modernity to postmodernity occurred in the mid-twentieth century, with the end of the Second World War, and in particular the disclosure of the facts of the Holocaust, marking the epochal border, the moment when what Jean-François Lyotard terms the 'grand narratives' of Enlightenment modernity were no longer sustainable (see Lyotard 1984). Robert Eaglestone, for instance, argues that 'postmodernism in the West begins with thinking about the Holocaust' (Eaglestone 2004: 2). As we shall see, however, when one considers their respective relations to the concept of nihilism, any clear distinction between modernity and postmodernity, or between modernism and postmodernism, is rendered highly problematic.

Modernism

As a critique of modernity, Horkheimer and Adorno's *Dialectic of Enlightenment* belongs to a tradition of philosophical modernism the origins of which Griffin locates in the second half of the nineteenth century, after the various failed revolutions of 1848. It was at that time that modernity, which had until then generally been seen as progressive, moving in the direction of an ever more enlightened future, began to be seen as a failed project (see Griffin 2007: 45). And this sense of failure was only reinforced by the experience of the Franco-Prussian War of 1870–1. While this view of the modernist reaction to modernity tends to sideline romanticism as the first major critique of modernity, and thus to obscure to some extent the connections between romanticism and modernism, Griffin is arguably right to claim that, with the publication of Nietzsche's *Birth of Tragedy* in 1872, the modernist critique of modernity goes 'beyond the sphere of aestheticism and contemplative philosophy to the realm of cultural criticism and metapolitics, the antechamber of social and political action' (94). This turn to metapolitics does not leave the aesthetic behind, however, since Nietzsche's critique of the modern age privileges art as 'the highest task and truly metaphysical activity of this life', and insists that 'it is only as an *aesthetic phenomenon* that existence and the world are eternally *justified*' (Nietzsche 1967: 31–2, 56). In the aftermath of the First World War, and in no small part on account of the growing influence of Nietzsche in Western Europe, the urgency of the modernist critique of modernity increased considerably across the philosophical, political, and aesthetic fields. The sense of modernity as desacralization, or as what Max Weber termed 'disenchantment' (*Entzauberung*), resulted in the interwar years in a range of resacralizing modernist projects, ranging from the various European avant-garde movements, to the fundamental ontology of Martin Heidegger, to Italian fascism and German Nazism.

As for aesthetic modernism, 1857 stands out as a decisive year, since it saw the publication of both Charles Baudelaire's *Flowers of Evil* and Gustave Flaubert's *Madame Bovary*. Both of these works were generated in part by what Jameson describes as a 'new and unadorned experience of time' – that is, time conceived both in its measurability (the working day) and as 'the deep bottomless vegetative time of Being itself, no longer draped or covered with myth or inherited religion' (Jameson 1994: 44). Following in Walter Benjamin's footsteps, Griffin identifies Baudelaire as 'one of the first Europeans to combine the description of modernity as a world that has lost its ordering principle and mythic centre – a world of "the transient, the fleeting, the contingent" – with the recognition that the artist's task is to wrest from it a sense of transcendence: "the eternal, the immutable"' (Griffin 2007: 92). As we shall see in the second part of this book, the privilege accorded by Baudelaire to the aesthetic in the struggle against the desacralizing tendencies of modernity recurs in various forms in philosophical, political, and aesthetic modernism in the first half of the twentieth century, and above all in conceptions of modernity as nihilist.

Although there are undoubtedly many reactionary and anti-modern elements in specific manifestations of philosophical, political, and aesthetic modernism, conceiving of modernism as a reaction against modernity does not mean that the former has necessarily to be understood as essentially reactionary in nature. Modernism as a whole is a highly complex phenomenon, combining progressive and reactionary elements and being orientated both to the past and to the future, while also insisting upon a new sense of the present. Furthermore, there are modernisms of the Left and of the Right, the differences between them often being far from easy to identify, and certainly more complex than is suggested by Walter Benjamin's distinction in his celebrated essay 'The Work of Art in the Age of Its Technological Reproducibility' (1936–9) between the aestheticization of politics (fascism) and the politicization of art (communism) (see Benjamin 2003: 207).

While he sees all forms of modernism as sharing a critical attitude to modernity, Griffin distinguishes between two principal modes: programmatic and epiphanic modernism. Whereas programmatic modernism looks to remake the world, epiphanic modernism withdraws from it. As Griffin puts it, in programmatic modernism 'the rejection of Modernity expresses itself as a mission to change society, to inaugurate a new epoch, to start time anew. It is a modernism that lends itself to the rhetoric of manifestos and declarations' (2). Epiphanic modernism, on the other hand, is characterized by 'the cultivation of special moments in which there is *Aufbruch* of a purely inner, spiritual kind with no revolutionary, epoch-making designs on "creating a new world"' (62). Put in the temporal terminology used by Frank Kermode in *The Sense of an Ending* (1967), these epiphanic moments are those 'in which the soul-destroying *chronos* of "waiting time" magically gives way to *kairos*, "a point in time filled with significance, charged with meaning derived from its relation to the end"' (Griffin 2007: 63). Griffin identifies Nietzsche as the paradigmatic figure in programmatic modernism, and Franz Kafka as Nietzsche's counterpart in epiphanic modernism. While this disposition of roles might seem to suggest that programmatic modernism finds its fullest expression in philosophy, and epiphanic modernism in art, the history of modernism suggests no such neat symmetry.

The distinction between two primary modes of modernist response to modernity can certainly help to clarify important differences in approach to the perceived nihilism of modernity, but the too rigorous application of the programmatic/epiphanic distinction leads inevitably to distortions. As Griffin points out, there are modernists who pass from the programmatic to the epiphanic mode, an example being the German poet Gottfried Benn, who in 1933 committed himself to Nazism and, after having been deeply troubled by Hitler's murderous purge of certain elements in the Party on the so-called Night of the Long Knives in the summer of 1934, 'returned to the safe haven of apolitical poetry' (65). Other major figures whose trajectory arguably carries them from programmatic to epiphanic modernism include Ernst Jünger and Martin Heidegger, both of whom came to take positions that, while not apolitical, were less obviously aligned with the extreme Right in the years immediately following the collapse of the Third Reich. Beyond such shifts from one mode to another, however, there are also instances of modernism where the distinction between the programmatic and the epiphanic does not hold, and, as we shall see, there are moments when this is the case in the work of both Nietzsche and Kafka.

Within the sphere of philosophical modernism, Nietzsche is paradigmatic above all because his critique of modernity exerts an influence greater than that of any other philosopher of the period, and because it was he who deployed the concept of nihilism to capture the essence of modernity. There is no small irony in the fact that this influence was owing principally to the posthumous publication of *The Will to Power* (1901; enlarged edition 1906), a work that bears no resemblance to the one that Nietzsche announced under that title, but which his editors put together from material in his 1880s notebooks and the first part of which was given the title 'European Nihilism'. Key figures in philosophical and aesthetic modernism whose critiques of modernity are profoundly marked by Nietzsche's diagnosis of modernity as nihilist include, to name but a few, Hugo Ball, Gottfried Benn, Ernst Jünger, Martin Heidegger, Wyndham Lewis, Theodor Adorno, Albert Camus, Maurice Blanchot, and E. M. Cioran. If Nietzsche's influence on these figures is evident in their conception of modernity as nihilist, it is no less evident in their privileging of the aesthetic, in one form or another, as what, in a May–June 1888 notebook entry that was included in *The Will to Power*, Nietzsche terms the 'only superior counterforce' to nihilism (Nietzsche 1968: 452, 1999b: xiii. 521).

What is it about art that it should be accorded this privilege in the modernist struggle against nihilism? And can a clear distinction be made between the turn to art and the turn to myth in philosophical, political, and aesthetic modernism? It has often been claimed that the modernist critique of modernity is essentially a critique of *logos* in the name of *muthos*, and that just such a turn to myth is to be found in Nietzsche's championing of the Dionysian. Myth here would serve to bestow meaning and order on experience, to effect a transcendence of the temporal as meaningless time (*chronos*). Guy Debord, for instance, sees fascism as 'a violent resurrection of *myth* which demands participation in a community defined by archaic pseudo-values: race, blood, the leader' (Debord 1983: section 109). Griffin maintains, however, that fascism has also to be seen as modernist in its orientation towards the future; that is, in Germany, the dream of a new, thousand-year Reich characterized by health, vitality, strength, racial purity, and national unity. Fascism, he argues, 'expresses a quintessentially primordial human drive to resolve the unprecedented socio-political and *nomic* crisis through which European history was passing after the First World War by constructing a new order which would provide "healthy" Italians and Germans with a new homeland, both material and mythic' (96). As a politically modernist response to modernity, fascism evidently combines the regressive and the futural. Indeed, modernism in its philosophical, political, and aesthetic forms is arguably characterized by just such a combination of opposing drives, and is never purely reactionary or progressive, nostalgic or forward-looking.

Furthermore, it would be a mistake to see the modernist turn to myth as necessarily identical with the privileging of art. One of the most important recent debates concerning Heidegger's thought, in which the concept of nihilism plays a decisive role from the mid-1930s onwards, is whether his understanding of art collapses the distinction between art and myth. According to Philippe Lacoue-Labarthe, for instance, Heidegger's readings of the poetry of Friedrich Hölderlin and Georg Trakl are 'remythologizations' in which the essence of poetry (*Dichtung*) becomes *muthos* (see Lacoue-Labarthe 2007: 33). One has also to take account here of Kermode's distinction between fiction and myth. According to Kermode, 'Myth operates within the diagrams of ritual, which presupposes total and adequate explanations of things as they are and were; it is a sequence of radically unchangeable gestures. Fictions are for finding things out, and they change as the needs of sense-making change. Myths are the agents of stability, fictions are the agents of change. Myths call for absolute, fictions for conditional assent' (Kermode 1967: 39). Crucially, fictions 'can degenerate into myths whenever they are not consciously held to be fictive', and, according to Kermode, anti-Semitism is just such a 'degenerate fiction' (39). The distinction between fiction and myth is, however, scarcely a hard and fast one, and the slippage from the former to the latter is apparent in the work of a large number of modernists on the political Right, including Richard Wagner, Ernst Jünger, Pierre Drieu La Rochelle, Ezra Pound, and Wyndham Lewis.

Nihilism

Within the sphere of aesthetic modernism, the concept of nihilism has been deployed both by the champions of modernist art, including some of the artists themselves, and by those critical of aesthetic modernism. Adorno is among the most influential champions of 'radically darkened' modernist art as the only fitting response to the nihilism of a modernity governed by the principle of identity, a nihilism that leads ultimately to genocide as the 'absolute integration' (Adorno 1973: 362). In direct contrast, Georg Lukács attacks both the avant-garde and high modernism as themselves nihilist. On the one hand, then, aesthetic modernism is seen, and sees itself, as the counterforce to the nihilism of modernity, while on the other hand it is identified as the embodiment of nihilism. And this situation is complicated further by some champions of the avant-garde celebrating that movement for its nihilism – as, for instance, Marcel Duchamp does when reflecting on Dada. In the interwar years in particular, one finds recurrent appeals to a nihilism that will clear the ground for a new cultural beginning.

In the extensive critical literature on modernism and modernity, the term 'nihilism' is generally used in a manner that implies its meaning is clear and circumscribable, and that no labour of interpretation or conceptual genealogy is required. To give just a few salient examples of this tendency: in Malcolm Bradbury and James McFarlane's widely read collection of essays *Modernism: A Guide to European Literature 1890–1930* (1976), modernism is described as being 'an extraordinary compound of the futuristic and the nihilistic' (Bradbury and McFarlane 1976: 46); in *Modernism* (2000), Peter Childs asserts that modernism combines 'nihilism and fanatical enthusiasm' (Childs 2000: 17); and in *Modernism and Fascism*, Griffin repeatedly uses the term in relation to both modernity and modernism. On the one hand, this reliance upon the term seems fully justified, given its repeated use by many of the most influential figures in philosophical, political, and aesthetic modernism, including Bourget, Nietzsche, Spengler, Jünger, Kafka, Heidegger, Hitler, Benjamin, Adorno, Camus, Cioran, and Blanchot. What such a reliance tends to leave out of account, however, is that, with each major deployment of the concept of nihilism in the modernist critique of modernity, its meaning shifts. As Jean-Pierre Faye puts it in *L'Histoire cachée du nihilisme* (2008), nihilism is a 'variable' (Faye and Cohen-Halimi 2008: 295). In short, there is no nihilism *as such*.

With this variability in mind, one of the principal aims of the present book is to chart the nature and the implications of some of the most important redeterminations of the concept of nihilism within the spheres of philosophical and aesthetic modernism. Since its introduction into the discourse on modernity at the time of the French Revolution, targets for the charge of nihilism have included atheism, Christianity, Judaism, rationality, metaphysics, ontology, transcendental idealism, logocentrism, deconstruction, technology, democracy, Nazism, fascism, socialism, bolshevism, humanism, and anti-humanism. It is one of the main methodological contentions of the present book that the concept of nihilism – which, as we shall see, lies at the very heart of the modernist critique of modernity – should be approached within the context of specific discourses. While it is undoubtedly owing to the influence of Nietzsche, and above all to the posthumous publication of *The Will to Power*, that the concept of nihilism occupies a central place within the modernist critique of modernity in the interwar years in particular, and while Nietzsche's influence remains the most important within the history of the thinking of nihilism in the postmodernist era, his determinations of nihilism – for there are more than one – have to be distinguished from those of many of the major figures in the history of the thinking of nihilism in the twentieth and twenty-first centuries. To repeat one of the underlying arguments of the present book: there is no nihilism *as such*; there are only specific deployments of the term, each of which has to be considered in its specificity, which means in its discursive context, including its relation to earlier determinations.

Some recent works on nihilism offer suggestive discriminations between various forms of nihilism. For instance, in *The Specter of the Absurd: Sources and Criticisms of Modern Nihilism* (1988), Donald A. Crosby distinguishes five distinct kinds: political, moral, epistemological, cosmic, and existential. Political nihilism he sees as embodied by the Russian nihilists of the second half of the nineteenth century. Moral nihilism takes three forms: amoralism ('the rejection of all moral principles and the determination to live without morality altogether'); moral subjectivism ('the theory that moral judgements are purely individual and arbitrary and admit of no rational justification or criticism'; and egoism ('the view that the sole obligation of any individual is to himself' (Crosby 1988: 11). Crosby finds examples of amoralism in the outlook of the character Wolf Larson in Jack London's novel *The Sea Wolf* (1904), moral subjectivism in the chapter on 'Science and Ethics' in Bertrand Russell's *Religion and Science* (1935), and egoism in Max Stirner's *The Ego and Its Own* (1845). Epistemological nihilism takes two main forms: 'The first makes claims to truth entirely relative to particular individuals or groups, while the second holds semantic intelligibility to be entirely relative to self-contained, incommensurable conceptual schemes' (Crosby 1988: 18). Crosby identifies Nietzsche as the prime epistemological nihilist, with another important figure being Fritz Mauthner, who in his *Contributions to a Critique of Language* (1901–2) proposes a radical linguistic relativism. Cosmic nihilism 'asserts the meaninglessness of the cosmos, either in the absolute sense of denying it any intelligibility or knowable structure at all, or in the relative sense of denying that it gives any place or support to the kinds of valuative and existential meanings to which human beings aspire' (Crosby 1988: 26). Crosby gives Schopenhauer, Nietzsche, and Russell as examples of major philosophers who articulate forms of such nihilism. Lastly, existential nihilism 'judges human existence to be pointless and absurd. It leads nowhere and adds up to nothing' (30). Crosby finds the mood of existential nihilism most eloquently expressed in Macbeth's speech when he realizes that he will inevitably be defeated by the forces assembled by Malcolm: 'Life's but a walking shadow; a poor player, / That struts and frets his hour upon the stage, / And then is heard no more; it is a tale / Told by an idiot, full of sound and fury, / Signifying nothing.' Other writers in whom existential nihilism finds expression would include Schopenhauer, Tolstoy, and Camus. Having distinguished these various forms of nihilism, Crosby goes on to identify what unites them, arguing that, whichever form it takes, the term 'nihilism' always implies negation or denial and that each type of nihilism denies a specific aspect of human life:

> *Political nihilism* negates the political structures within which life is currently lived, as well as the social and cultural outlooks that inform these structures. It has little or no vision of constructive alternatives or of how to achieve them. *Moral nihilism* denies the sense of moral obligation, the objectivity of moral principles, or the moral viewpoint. *Epistemological nihilism* denies that there can be anything like truths or meanings not strictly confined within, or wholly relative to, a single individual group, or conceptual scheme. *Cosmic nihilism* disavows intelligibility or value in nature, seeing it as indifferent or hostile to fundamental human concerns. *Existential nihilism* negates the meaning of life. (35)

Crosby considers all but the first of these forms of nihilism as philosophical in nature, and in his own critique of nihilism takes existential nihilism to be the primary form.

In *The Banalization of Nihilism* (1992), Karen L. Carr offers a slightly different breakdown of the concept, distinguishing between epistemological, aletheiological, metaphysical or ontological, ethical or moral, and existential or axiological nihilism (Carr 1992: 17–18). Carr departs from Crosby principally in her distinction between the denial of the possibility of knowledge and the denial of the reality of truth (both of which fall under epistemological nihilism in Crosby's schema), and in her conception of metaphysical nihilism as the denial of any independently existing world (this idea not being included by Crosby). Neither Crosby nor Carr considers nihilism in direct relation to the histories of modernity and modernism, however, and both set out to critique positions that they consider to be nihilist. In this, they join a tradition that extends back to Nietzsche and that is characterized by its attempt to diagnose the nature and causes of nihilism, and to point the way towards its overcoming.

The guest and its counterforce

The present work is distinct from previous studies of nihilism in a number of key respects. It aims to provide a critical history of the thinking of nihilism in its relation to both modernity and modernism through an analysis of explicit deployments of the term 'nihilism' in the fields of philosophical and aesthetic modernism. It is neither a critique nor a championing of nihilism, and thereby takes its distance from both the modernist and the postmodernist approaches to nihilism with which it engages. Chapters 1 and 2 focus on some of the major deployments of the concept of nihilism in philosophical modernism. Chapters 3 and 4 focus on the deployment of the concept of nihilism in aesthetic modernism and in the critical literature thereon. Chapter 5 focuses on the distinction between modernism and postmodernism in their respective engagements with the concept of nihilism. As Griffin observes, Nietzsche is undoubtedly the 'outstanding incarnation' of philosophical modernism, his importance lying in no small part in his having made of nihilism arguably the key philosophical concept of the first half of the twentieth century, together with his privileging of the aesthetic as the 'only superior counterforce' to nihilism. For these reasons, Nietzsche figures not only in the consideration of philosophical modernism, but also in that of aesthetic modernism. As regards the latter, Kafka occupies a central position because he is arguably, as Griffin puts it, 'the archetypal literary modernist' (375), his importance in the present context owing to his oeuvre's having been the one around which debates concerning the relation between nihilism and the aesthetic have tended to centre.

Ideally, a consideration of aesthetic modernism in relation to the concept of nihilism would include not only literature but also the other arts, especially painting and film. The films of the Italian director Michelangelo Antonioni, especially *L'Avventura* (1960) and *La Notte* (1961), are just one example of a body of work the critical commentary on which has raised the question of nihilism. For the most part, however, considerations of the relationship between aesthetic modernism and nihilism have focused principally on the literary, and it is to take account of this that the present work limits itself by and large to that realm. If a separate analysis of political modernism in its relation to nihilism is not offered here, this is in part because the concept of nihilism is from the outset political, and both philosophical and aesthetic modernism are political precisely in their engagement with the concept of nihilism.

It cannot be over-emphasized that the present book is not aligned with any of the determinations of nihilism considered therein, but rather considers the nature and implications of these determinations for a thinking of both modernism and modernity. If there is a lesson to be learned from the (selective) history outlined in this book, it is that none of these deployments takes full account of nihilism as what Nietzsche describes in an entry in his autumn 1885 to autumn 1886 notebook as 'this uncanniest of all guests' (*dieser unheimlichste aller Gäste*) (Nietzsche 1968: 7, 1999b: xii. 125). Indeed, the history of the deployment of the concept of nihilism in relation to both modernity and modernism – and to postmodernity and postmodernism – is characterized above all by this failure to make nihilism serve a discourse of critique. What unites almost all of the figures considered in the present work is not only their deployment of the concept of nihilism as part of a critique of modernity, but also their commitment to an overcoming of nihilism, and, crucially, their privileging of the aesthetic in one form or another as (in Nietzsche's words) the 'only superior counterforce' to nihilism.

In an earlier work (2008), I sought to chart the manner in which nihilism, in its uncanniness, returns to haunt the discourses of those philosophers and literary theorists who would make use of it in the interests of critique. In places, the present work returns to arguments made in that book, especially in the sections on philosophical modernism and postmodernism. The detailed analyses required to demonstrate the uncanny return of nihilism, and the manner in which it resists the mastery of those who would deploy it, could not be included in a short monograph the range of which is so extensive – aiming, as it does, to cover a period from the late eighteenth century to the present day, and including both philosophical and aesthetic modernism. Nonetheless, the uncanniness of nihilism can be seen in the kinds of reversal that are charted in the chapters that follow, reversals that enable nihilism to be directed as a charge against both atheism and Christianity, socialism and fascism, humanism and anti-humanism, the East and the West, logocentrism and deconstruction, the appeal to *logos* and the appeal to *muthos*, modernity and modernism, and, indeed, nihilism itself, for, as we shall see, one form of nihilism comes to be turned against another, even if the distinction between them is anything but clear.

In contrast to my earlier work on the concept of nihilism, the focus in the present book is on the function of that concept within a modernist critique of modernity that is at once aesthetic, political, and philosophical. Particular attention is given here to the discursive and national migrations of the concept: from mid-nineteenth-century Russian political discourse (Pisarev, Nechaev) to its appropriation by Nietzsche in the 1880s, by major figures in the European avant-garde in the early decades of the twentieth century (Hugo Ball, Marcel Duchamp), by conservative revolutionaries in Germany in the interwar years (Ernst Jünger, Martin Heidegger), by 'high' modernists (Kafka, Lawrence, Lewis), by major figures in the fields of philosophy and literature in France in the 1940s (Camus, Blanchot, Cioran), by influential commentators on modernist art in the 1950s (Lukács, Adorno), and by philosophers whose work tends to be aligned with the postmodern turn (Derrida, Baudrillard, Vattimo). For each phase in the concept's migration, it has been necessary, for obvious reasons, to be selective, and I have focused on figures within philosophical and aesthetic modernism who may be considered paradigmatic.

Part I

PHILOSOPHICAL MODERNISM AND NIHILISM

1

FROM THE FRENCH REVOLUTION TO NIETZSCHE

'The first great historical experience of the void'

In its appeal to reason and to the rights of freedom and equality for the individual conceived as bourgeois citizen, the French Revolution has often been seen as the inauguration of political *modernity*. At the same time, however, in its theorization as a radical break with the past (in the form of the ancien régime), that same revolution may also be seen as the inauguration of political *modernism* – in short, as what Andrew Gibson describes as the 'first great historical experience of the void underlying established structures, and therefore of the possibility of the tabula rasa and radical transformation' (Gibson 2006: 257). In what is no doubt more than a mere philological coincidence, it is in this co-foundation of political modernity and political modernism that the first deployment of the term 'nihilism' within political discourse takes place.

In a short text published on 26 December 1793, Jean-Baptiste du Val-de-Grâce, baron de Cloots (1755–94), a Prussian writing in French under the *nom de plume* Anarcharsis Cloots, argued that the founding of a republic of 'sovereign people' is possible only through the negation of all relation to the *theos*. Hence, not only must the new French republic avoid *theism*, it must also avoid *atheism*, since an atheist republic would remain determined, albeit negatively, by a theological relation. A republic beyond both theism and atheism is, he claims, *nihilist*: 'The republic of the rights of man is strictly speaking neither theist nor atheist, but nihilist' (Cloots in Biaggi 1998: 57; my translation). Governed solely by reason, political nihilism thus defined tolerates no spectre of the divine. Unsurprisingly, for having championed such a nihilism, Cloots soon found himself among the victims of the Terror. On 24 March 1794, he was guillotined together with the Hébertists, who were similarly opposed to Robespierre's Cult of the Supreme Being.

According to Roger Griffin, the Terror is the first instance within the political sphere of what, in the 1880s, Friedrich Nietzsche will term 'active nihilism' – in short, the first act of political modernism. As Griffin puts it: 'At the height of its sustained orgy of creative destruction, the French Revolution could already hatch plans for the purification of society expressed in a discourse that directly anticipates the twentieth-century discourse of biopolitics and social hygiene' (Griffin 2007: 183–4). The telling irony here would thus be that the victims of this active nihilism included the first theorist of political nihilism, and that two forms of nihilism were set against each other on the occasion of this first deployment of the term. At its very inception, then, philosophico-political nihilism was at war with itself, exhibiting what Jacques Derrida, in his work on the foundations of sovereignty and the nation-state, has described as 'autoimmunity' (see, for instance, Derrida 2005b: 123).

Nihilism as defined by Cloots operates as the governing principle of an absolutely secular sovereign republic, founded on the voiding of any authority beyond itself. The radicality of this political negation of the theological, which must go beyond that of the privative 'a-' of atheism, is such that it would leave no trace of that which it negates. Whereas later forms of political modernism, and in particular fascism, present themselves as *countering* the nihilism of modernity, in this first deployment of the term within philosophico-political discourse political *modernism* explicitly presents itself as political *nihilism*.

Ironically, despite his insistence on political nihilism's being defined by its break with the theological, through his deployment of the term 'nihilism' Cloots in fact ties his own political discourse to the theological insofar as that term had previously been deployed in relation to a specific religious doctrine. Three decades before the publication of Cloots's text, in the first volume of his *History of the University of Paris, from Its Origin to the Year 1600* (1761), Jean-Baptiste-Louis Crevier (1693–1765) made use of the term when referring to a heresy originating in the work of the twelfth-century scholastic theologian Peter Lombard. According to Crevier, in the fourth book of his *Libri Sententiarum* (*Books of Sentences*) Lombard argues that '*in as much as he is a man Jesus Christ is not something*, or, in other words, is *nothing*. This proposition is scandalous, and yet some of his disciples supported it and formed the heresy of the *nihilists*, as it was called' (Crevier 1761: 206; Crevier's emphasis; my translation).

Theological in its origins, then, the concept of nihilism very soon became not simply political but anti-theologically political. First, in the Lombardian heresy, it names the nothingness of Christ's humanity, and then, with Cloots, the 'nullity or invalidity of cults' (*nullité des cultes*) (Cloots in Biaggi 1998: 640). The term 'nihilism' appears to originate in a theological discourse, to migrate first to the political and then, as we shall see, to the philosophical and ultimately to the aesthetic. However, the history of nihilism, its many migrations and counter-migrations, its many reversals and counter-reversals, will in fact come to trouble all watertight conceptual borders between the theological, the political, the philosophical, and the aesthetic. The disturbance of these borders is already apparent in Cloots's discourse, which may be seen as a failed attempt to establish nihilism as a purely political concept. A second disturbance of discursive borders occurs with the term's migration six years later, this time from the political to the philosophical. In fact, this second migration returns the concept of nihilism both to the philosophical-theological – more precisely, to the distinction between faith and reason – and to the political. For the philosophy that is branded as nihilism is rooted in the Kantian revolution, which is to say the philosophical revolution often seen to be a key factor in the political revolution of 1789.

'The Jews of speculative reason': Jacobi's letter to Fichte

According to both Martin Heidegger and Theodor Adorno – two of the most significant twentieth-century philosophers to have engaged with the concept of nihilism, although from opposing political positions – the term's first properly *philosophical* deployment occurs in an Open Letter dated 21 March 1799 (and published in the autumn of that year) from the German philosopher and novelist Friedrich Heinrich Jacobi (1743–1819) to Johann Gottlieb Fichte, whom Jacobi identifies in this letter as being 'among the Jews of speculative reason their king' (Jacobi 1987: 122–3). The letter to Fichte forms part of a more general attack by Jacobi on Kantian idealism. What distinguishes this particular attack from all others, however, is that here Fichte's idealism is denounced as nihilism (*Nihilismus*) on account of its annihilation of anything beyond the transcendental ego. Strictly speaking, such a philosophy is science (*Wissenschaft*) in the sense that it transforms everything that is into nothing, leaving only 'a spirit so pure that it cannot itself exist in this its purity but can only generate everything' (126). Scientific comprehension entails the negation of that which exists 'in itself' so that it can become 'completely our own creation subjectively' (127). What is lost in such philosophical nihilism is precisely that which lies beyond the grasp of *Wissenschaft*, namely 'the true', 'goodness', or 'God'. These, according to Jacobi, simply cannot be a 'creation of human invention', and are radically other to the human as a purely reasoning being (130). By 'the true' here, Jacobi means 'something which is prior to and outside of knowing; which first gives knowing and the capacity for knowing, for reason, its value' (131). And 'God', as he puts it in the preface to the published version of the letter, 'cannot be known, but only believed. A God who could be known would be no God at all' (121). God is that 'highest being above and outside of me', which I must come to accept 'instinctively' (132). Of the true, goodness, or God – and precisely on account of their radical alterity to the ego – the 'I' can have no knowledge but only a 'distant presentiment' (133). If one denies the existence of anything beyond the world of appearances, then those appearances themselves become 'abstract ghosts, appearances of Nothing' (131).

In countering the nihilist 'knowing of nothing' that would reduce God, goodness, and the true to nothing, Jacobi does not propose anything positive. Rather, he champions a 'non-philosophy' (*Unphilosophie*), a 'philosophy of not knowing', or 'science of ignorance' (136, 140). Paradoxically, then, while there can be no knowing of God, since any such knowing would be an annihilation of the object of knowledge, and thus nihilism as atheism, there can be not only a presentiment of God but also a *knowledge of not knowing*. In short, the limits of knowledge can themselves be known. For Jacobi, the border between the known and the unknowable remains secure, patrolled and maintained by knowledge.

Although both Heidegger and Adorno identify Jacobi's deployment of the term 'nihilism' as its first *philosophical* use, this deployment is in fact *theologico*-philosophical in nature, concerning as it does the reduction to nothing of God and thus belonging within the tradition of thinking nihilism inaugurated by Crevier and including Cloots. That said, if nihilism stands for atheism in both Cloots and Jacobi (for whom Kantian deism is a disguised atheism), Jacobi's deployment of the term involves a complete reversal of values. Whereas Cloots champions nihilism as a negation of the *theos* beyond that achieved in atheism, Jacobi denounces it. And this reversal occurs within the same general field of discourse, since the philosophy denounced as nihilist by Jacobi finds its origin in Kant, whose thought lies not only behind *philosophical* modernism but also behind *political* modernism insofar as the two will prove to be inseparable.

In Jacobi, then, one finds an attack not only on atheism disguised as deism, and on the absolute privilege accorded to reason, Logos, and science in the Enlightenment (*Aufklärung*), but also on philosophico-political modernity *as nihilism*. And, in this respect, he may be said to inaugurate an entire tradition, in which it is precisely modernity in its various forms (philosophical and political) that will be charged with nihilism. That Jacobi should see this nihilism as being embodied in the work of a philosopher whom he labels the king of the 'Jews of speculative reason' also marks the commencement of a tradition that will have consequences scarcely dreamed of in Jacobi's philosophy. In this tradition, which is at once theological, philosophical, political, and aesthetic, the 'Jew' becomes the paradigmatic figure of the nihilist. As we shall see, this tradition, too, will be marked by its conceptual and evaluative reversals.

'The passion for destruction': Russian nihilism

If Cloots's conception of nihilism is theologico-political in nature, and Jacobi's theologico-philosophical, the next major deployment of the term returns us to the political, although the discourse in which this deployment occurs is, for the first time, a literary one. This migration of the term 'nihilism' to the literary occurs in 1862. It was in the spring of that year that Victor Hugo published his novel *Les Misérables*, in which nihilism is defined as, and criticized for being, the reduction of the infinite to a 'concept of the mind', this definition recalling Jacobi's:

> To deny will to the infinite, – that is, to God, – is only possible by denying the infinite.

> Negation of the infinite leads straight to nihilism [*mène droit au nihilisme*]; everything becomes a 'concept of the mind'.
>
> With nihilism, no discussion is possible; for logical nihilism doubts the existence of its opponent, and is not sure of its own. It is possible, from its standpoint, that it is only a 'concept of the mind'. It does not perceive, however, that it admits everything which it has denied when it utters the word 'mind'.
>
> In brief, no path for thought is opened by a philosophy which makes everything end in the monosyllable, No.
>
> To No there is but one answer, – Yes.
>
> Nihilism has no compass.
>
> There is no such thing as nothingness. Zero does not exist. Everything is something. Nothing is nothing. (Hugo 1907: 242)

It is, however, another novel published in the same year that was responsible for the wider circulation of the concept of nihilism across Europe. In *Fathers and Sons*, the Russian writer Ivan Turgenev (1818–83) offers what is by no means an entirely damning representation of the 'Russian nihilist' in the form of Bazarov, a figure modelled, according to Turgenev, on a doctor whom he met on a train in Russia and whose ideological position is summed up early in the novel as follows: 'A nihilist is a person who does not take any principle for granted, however much that principle may be revered' (Turgenev 1975: 94). As Isaiah Berlin observes, 'The central topic of the novel is the confrontation of the old and the young, of liberals and radicals, traditional civilization and the new, harsh positivism which has no use for anything except what is needed by a rational man' (Berlin 1975: 26–7). Bazarov's brand of nihilism appeals to reason, empirical observation, experiment, and hard facts, and rejects anything that cannot be scientifically demonstrated, including religion, the arts, and liberal values. The aim of such a nihilism is not to construct anything, not to replace the existing order with a new one, but, as Bazarov puts it, simply to 'clear the ground'. For this reason, Bazarov's nihilism may be described as a form of political modernism.

Although the French writer Guy de Maupassant (in an article published in *Le Gaulois* in November 1880) claims that Turgenev was 'the inventor of the word "nihilist", the first to have drawn attention to this sect that is so powerful today, and who, so to speak, legally named it' (Maupassant in Biaggi 1998: 65; my translation), the term had in fact already been circulating within political discourse since the 1830s. That said, Turgenev was undoubtedly responsible for the term's becoming common currency both in Russia and in Western Europe, especially in France and Germany, and this wider circulation increased exponentially in the wake of the assassination of Tsar Alexander II in March 1881, and the sequence of other assassination attempts on European heads of state in the last two decades of the nineteenth century and the first decade of the twentieth, when the distinction between political nihilism and anarchism was blurred, to say the least. These included failed attempts on the lives of the German Emperors Wilhelm I (in 1883) and Wilhelm II (in 1900 and 1901), and the assassination of the French President Marie-François-Sadi Carnot in June 1894, and the Empress Elisabeth ('Sisi') of Austria in September 1898.

The sect to which the term 'nihilist' referred in the early 1860s was far from being a unified or coherent political movement; rather, it tended to be applied loosely to a new revolutionary generation made up of anarchists, socialists, and other political groupings in Russia calling not simply for the questioning of every principle but for a decisive break with the past through the destruction of the existing socio-political institutions and apparatuses. During his time at the University of Berlin in 1838–41, Turgenev had heard the word 'nihilist' being used of those on the Hegelian Left, including the future theorist of anarchism Mikhail Bakunin (1814–76), whom the novelist met in Berlin in 1840 and whose most fundamental principle is echoed in Bazarov's claim that what is required is not to found anything but simply to clear the ground.

Having begun to study German Idealism in 1836, and drawing in particular on Hegel's thinking of the negative in the *Science of Logic* (1811–16), Bakunin was already dreaming of what, in his 1842 essay 'The Reaction in Germany: A Fragment from a Frenchman', he describes as 'a new, vital, and life-creating revelation, a new heaven and a new earth, a young and magnificent world in which all present discords will resolve themselves into harmonious unity'. The well-known final paragraph of this essay constitutes nothing less than a call to nihilist arms: 'Let us […] trust the eternal Spirit which destroys and annihilates only because it is the unfathomable and eternally creative source of all life. The passion for destruction is a creative passion, too' (Bakunin 1973: 40, 58). The modernism of Bakunin's political thinking is evident here precisely in this dream of a radical break with the past and an absolute renewal, the model for which he finds, significantly, in the French Revolution: 'Have you not read the mysterious and awesome words, *Liberté, Égalité* and *Fraternité* on the foreground of the Temple of Freedom erected by the Revolution? And do you not know and feel that these words intimate the complete annihilation of the present political and social world?' (55). It is this call for a *complete annihilation* that placed Bakunin at the centre of the various revolutionary movements in Russia that Turgenev labelled as nihilist, even if the figure of Bazarov in *Fathers and Sons* is not, like the protagonist of Turgenev's 1856 novel, *Rudin*, based specifically on Bakunin.

Even more extreme in his passion for destruction than Bakunin was Sergei Nechaev (1847–82), who became the model for another literary figure, Peter Verkhovensky in Fyodor Dostoevsky's 1871 novel, *Demons*, which Ronald Hingley has rightly described as 'the greatest work of imaginative literature devoted to Nihilism' (Hingley 1967: 46). In 1869, Nechaev published a pamphlet entitled *Catechism of a Revolutionary*, in which he preached absolute commitment to the destruction of the entire socio-political world as it currently existed: 'We recognize', he declared, 'no other authority but the work of extermination, but we admit that the forms in which this activity will show itself will be extremely varied – poison, the knife, the rope etc. In this struggle, revolution sanctifies everything alike' (quoted in Carr 1975: 379). Nechaev was certainly prepared to act on this principle, being personally responsible for the political murder of a student named Ivanov at the Petrovsky Agricultural Academy in Moscow in November 1869. It was this murder that particularly struck Dostoevsky, leading him to make nihilism the principal focus of *Demons*.

The reception in Russia of Turgenev's literary portrayal of the nihilist in *Fathers and Sons* lacked any consensus, some critics arguing that it was sympathetic to the nihilists, others that it was critical of them, and others again that it was ambiguous in its attitude. In contrast, Dostoevsky's literary depiction of the nihilists in *Demons* is no less damning than his various comments on them in his correspondence. These comments reveal that, for Dostoevsky, the essence of nihilism lies (as it will for Heidegger in the 1930s) in unrootedness, a detachment from the homeland, the 'native soil'. Furthermore, in his correspondence of the 1860s and 1870s, Dostoevsky makes no distinction between nihilists and socialists, and in this he anticipates right-wing appropriations of the term in Western Europe in the post-First World War period. In a letter of 25 April 1866 to M. N. Katkov, for instance, Dostoevsky declares:

> The doctrine that 'everything should be shaken up par les quatre coins de la nappe, so that at least there may be a tabula rasa for action' – such a doctrine needs no roots. All nihilists are socialists. Socialism (especially in its Russian variety) specifically requires that all links should be cut. Why, they are absolutely convinced that, given a tabula rasa, they could at once build a paradise on it. (Dostoevsky 1987: 229)

In a letter of 25 March–6 April 1870 to A. N. Maikov, he writes:

> Nihilism isn't even worth talking about. Wait until the upper layer, which has cut itself loose from the Russian soil, rots through and through. And you know, it seems to me sometimes that many of those young scoundrels, those decaying youths, eventually will become real, solid pochvenniki, true Russians deeply attached to their native soil. As to the rest, let them rot away. They will be struck dumb by paralysis. Ah, but what a lot of scoundrels they still are! (333)

And, in a letter of 1 March 1874 to V. P. Meshchersky, he refers to '*the nihilist scum*' (386; Dostoevsky's emphasis).

It is, however, in a letter of 29 August 1878 to V. F. Putsykovich that Dostoevsky makes a claim that will recur in the later history of the deployment of the term 'nihilism' as part of a discourse that goes far beyond even Nechaev's justification of extermination. We have seen that, already in Jacobi, a connection is made between nihilism and the Jews when he identifies Fichte as the king of the 'Jews of speculative reason'. Dostoevsky goes much further in the letter to Putsykovich:

> Incidentally, when will they finally realize how much the Yids (by my own observation) and perhaps the Poles are behind this nihilist business. There were a bunch of Yids involved in the Kazan Square incident, and then it was Yids in the Odessa incident. Odessa, the city of the Yids, is the center of our militant socialism. In Europe, it's the same situation: the Yids are terribly active in socialism, and I'm not speaking now about the Lassalles and the Karl Marxes. Understandably so: the Yid has everything to gain from every cataclysm and coup d'état, because it is he himself, status in statu, who constitutes his own community, which is unshakable and only gains from anything that serves to undermine non-Yid society. (461)

It is, then, precisely in what he takes to be the *unrootedness* of the Jews, their forming a community that is independent of any national homeland, that they are able to incarnate a nihilism directed against the existing socio-political order and its institutions. Dostoevsky's claim here also returns us to the theological origins of the term in the Lombardian heresy denying the humanity of Christ. For, just as that 'nihilist' heresy denied the incarnation, so Dostoevsky finds nihilism in a people whose religion denies the Christian Messiah. The politically reactionary, deeply Slavophile Dostoevsky here anticipates by some decades the connection made between nihilism, socialism, and 'international Jewry' by the ideologues of Nazism.

'The uncanniest of all guests': Nietzsche and the radical inflation of nihilism

If Dostoevsky's *Demons* proved crucial to the later thinking of nihilism, then this was above all on account of its impact upon Friedrich Nietzsche (1844–1900), who came across Dostoevsky's work in early 1887. Roger Griffin identifies Nietzsche as 'one of the outstanding figures of the revolt against Modernity-as-Decadence' (Griffin 2007: 58), but Nietzsche is arguably not merely one among others. Indeed, he is *the* outstanding figure in that revolt insofar as it involves the deployment of the concept of nihilism, both for his radical reconceptualization of nihilism and for the scarcely calculable influence on twentieth-century thought of his cultural critique of modernity, within which the concept of nihilism came to play such a decisive role. What Griffin identifies as Nietzsche's 'programmatic modernism' – which is to say, a modernism that aims at a complete renewal through what Nietzsche terms the 'revaluation of all values' – situates the concept of nihilism at the very heart of philosophico-political modernism's response to modernity. As we shall see in the next chapter, this centrality is nowhere more obvious than in the work of another major German philosopher, Martin Heidegger.

According to Keith Ansell-Pearson, Nietzsche sees Western civilization as 'caught in the grip of debilitating and demoralizing nihilism in which our most fundamental conceptions of the world are no longer tenable and believable. Nihilism is thus a condition which affects the metaphysical and moral languages through which we fabricate an understanding of the world and on which we base our acting in the world' (Ansell-Pearson 1994: 7). While this is a fair overview of Nietzsche's thinking of nihilism, it is nonetheless important to distinguish between two historical phases in the reception of Nietzsche's thought. These phases are tied to two very different kinds of corpus, and relate directly to the place occupied by the concept of nihilism within his critique of modernity.

The first of these two phases is the reception of those works published by Nietzsche between 1872, the year in which his first book, *The Birth of Tragedy out of the Spirit of Music*, appeared, and 1887, when *On the Genealogy of Morals* was published, the latter being the text in which Nietzsche declares that he intends to write a work entitled 'On the History of European Nihilism', which 'will be contained in a work in progress: *The Will to Power: Attempt at a Revaluation of All Values*' (Nietzsche 1989: 159–60). As Alfred Weber observed in 1945 – when Europe, and above all the Germany from which Weber was writing, lay in ruins as a result of what many took to be a political nihilism inspired by Nietzsche – this Nietzsche of the works published between 1872 and 1887 is 'the man who influenced the young generation of the nineties, a period rich in spiritual contradictions that brings the Nineteenth Century to a close and heralds the first European and world catastrophe' (Weber 1947: 116). Heidegger makes a similar point at the beginning of his 1936–7 lecture series on Nietzsche and 'The Will to Power as Art': 'for a long time it has been declaimed from chairs of philosophy in Germany that Nietzsche is not a rigorous thinker but a "poet-philosopher"' (Heidegger 1979: 5). The key work for those who championed Nietzsche as such a poet-philosopher was *Thus Spoke Zarathustra* (1883–5), in which, as it happens, the word 'nihilism' does not appear.

There is, however, another Nietzsche, namely the one who, in Weber's words, produced a 'different wave of influence that only reached its full height between the first and the second world wars. To understand this wave whose profound and – we must say bluntly – fateful effect was due to his popularization we have to bear in mind that his teachings only came to a practical head in the book that appeared posthumously at the beginning of the Twentieth Century, […] *The Will to Power*' (Weber 1947: 117). While the term 'nihilism' occurs repeatedly in the works of 1887–8, including *On the Genealogy of Morals*, *Twilight of the Idols*, *The Antichrist*, and *Ecce Homo* (the last three of these being written at breakneck speed in 1888, but published after Nietzsche's lapse into madness, in 1889, 1894, and 1908 respectively), it is only in *The Will to Power* that the concept of nihilism takes centre-stage in Nietzsche's critique of modernity. That *The Will to Power* as published in 1901, and then in a considerably enlarged edition in 1906, was not in fact a book written by Nietzsche, or even, as he claims, a 'work in progress', but rather a collection of fragments from his notebooks, is of considerable importance here. It is not Nietzsche himself, but the editors of his so-called 'Nachgelassene Werke' in the *Grossoktav* edition (1894–1904), who place nihilism at the forefront of his late thought, with major philosophical and political consequences for the twentieth century.

According to Elisabeth Kuhn, the texts in which Nietzsche first encountered the term 'nihilism' were neither philosophical nor political works but the French translations of Turgenev's novels *Fathers and Sons* (which he read in the summer of 1873) and *Virgin Soil*, together with Prosper Mérimée's published remarks on these texts (see Kuhn 1984: 262–3), and then the two volumes of Paul Bourget's *Essays in Contemporary Psychology* (1883 and 1885). Bourget uses the term *nihilisme* repeatedly of an entire movement in French literature from Baudelaire to Flaubert and Maupassant, a movement characterized, according to Bourget, by its 'pessimism', its 'misanthropy', and its 'world-weariness' (Bourget 1912: i. xxii). In order to grasp why there should have been a sudden explosion of the term 'nihilism' in Nietzsche's writings during the last two years of his productive life, however, one has to take account of his reading of Dostoevsky in 1887, and in particular of the novel *Demons*.

Dostoevsky's profound impact on Nietzsche is revealed in a letter of 23 February 1887 to Franz Overbeck, in which he writes, from Nice:

> I […] knew nothing about Dostoevski until a few weeks ago – uncultivated person that I am, reading no 'periodicals'! In the bookshop my hand just happened to come to rest on *L'Esprit souterrain*, a recent French translation (the same kind of chance made me light on Schopenhauer when I was twenty-one, and on Stendhal when I was thirty-five). The instinct of affinity (or what shall I call it?) spoke to me instantaneously – my joy was beyond bounds; not since my first encounter with Stendhal's *Rouge et Noir* have I known such joy. (Nietzsche 1969: 260–1)

Having begun with *Notes from Underground* (1864), which he read in French, as he had Turgenev's *Fathers and Sons*, Nietzsche proceeded to work his way through Dostoevsky's major novels. Of these, it is *Demons* that leaves the most significant traces in Nietzsche's notebooks, where, in 1887 – which Jean-Pierre Faye rightly dubs 'the *decisive year* of Nietzschean nihilism' (Faye and Cohen-Halimi 2008: 256; Faye's emphasis) – passages from the Verkhovensky–Stavrogin dialogues are copied down (in Notebook 11, November 1887–March 1888; e.g. Nietzsche 1999b: xiii. 147–50). These passages include Verkhovensky's declaration: 'I am a nihilist, but I love beauty' (*Ich bin Nihilist, aber ich liebe die Schönheit – je suis nihiliste, mais j'aime la beauté*) (Dostoevsky 2000: 419; Nietzsche 1999b: xiii. 149). This paradox at the heart of nihilism as characterized by Dostoevsky will find a distorted echo in Nietzsche's own thinking of nihilism.

As Joseph Frank observes, the target of Dostoevsky's critique of nihilism in *Demons* is above all Western rationalism, which Dostoevsky saw as 'inevitably leading to the replacement of the God-man Christ, with his morality of love, by the Man-god of egoism and power' (Frank in Dostoevsky 2000: xxvi). As the embodiment of 'rational egoism', Dostoevsky's nihilist is, then, anti-Christian. Nietzsche's notebooks certainly bear witness to his interest in the term 'nihilism', this interest also being made public in his claim in *On the Genealogy of Morals* that he intends to write a piece entitled 'On the History of European Nihilism'. Far from simply adopting Dostoevsky's, or indeed Turgenev's, conception of nihilism as anti-Christian, however, Nietzsche turns that conception on its head, and in so doing he breaks completely with Dostoevsky's identification of the Jew as the archetypal nihilist.

In the works written by Nietzsche for publication, the term 'nihilist' first appears in *Beyond Good and Evil: Prelude to a Philosophy of the Future* (1886), where it refers to those 'fanatics of conscience who would rather lie dying on an assured nothing than an uncertain something' (Nietzsche 2002: 11). The words 'nihilism' (*Nihilismus*) and 'nihilist' (*nihilistisch*) are then used repeatedly in the sequence of works written between 1887 and Nietzsche's collapse in January 1889, including *On the Genealogy of Morals*, where it appears more often than in any of his other published works, *Twilight of the Idols*, *The Antichrist*, and *Ecce Homo*. In these texts, the term 'nihilism' is almost always used in reference to religion, and in particular to Christianity. In *Ecce Homo*, for instance, both Christianity and Buddhism are described as 'nihilistic religions' (Nietzsche 2005: 16). Often, one has the sense that Nietzsche is simply using 'nihilist' as a synonym for 'decadent', and, in *Ecce Homo*, Christianity and Buddhism are also labelled 'religions of decadence'.

In addition to his use of the term in the new works written in 1887–8, Nietzsche also inscribed the concept of nihilism into his earlier works. For instance, he uses the word in the 'Attempt at a Self-Criticism' that prefaces the 1886 edition of *The Birth of Tragedy*, and the section on *The Birth of Tragedy* in *Ecce Homo* includes the definition of Christianity as 'nihilist' on account of its being neither Apollonian nor Dionysian, but rather a religion that '*negates* all *aesthetic* values' (Nietzsche 2005: 108; Nietzsche's emphasis). It is precisely this negation of aesthetic values that, according to Nietzsche, makes Christianity 'nihilistic in the deepest sense' (108). As his notes on nihilism of the same period reveal in much greater detail, Nietzsche sees both the origin and the essence of nihilism as lying in the moral interpretation of phenomena, this interpretation reaching its most extreme form in Christianity. For Nietzsche, the opposite of the moral interpretation is an aesthetic interpretation, and so it is that nihilism and art come to be placed in diametrical opposition to each other. But why is the moral interpretation of phenomena nihilist? And what are those 'aesthetic values' that are negated by this moral interpretation?

In an autumn 1887 notebook entry, Nietzsche poses the question 'What does nihilism mean?', and answers: '*That the highest values devaluate themselves*' (Nietzsche 1968: 9, 1999b: xii. 350; Nietzsche's emphasis). The values to which he is referring here are unity (*Einheit*), purpose or aim (*Zweck*), and truth (*Wahrheit*). These three highest values 'devaluate themselves' when it is realized that

> the overall character of existence may not be interpreted by means of the concept of 'aim', the concept of 'unity', or the concept of 'truth'. Existence has no goal or end; any comprehensive unity in the plurality of events is lacking; the character of existence is not 'true', is *false*. One simply lacks any reason for convincing oneself that there is a *true* world. (Nietzsche 1968: 13, 1999b: xiii. 48; Nietzsche's emphasis)

The phrase 'God is dead' is Nietzsche's well-known shorthand for this devaluation. What that phrase tends to obscure, however, is that it is not only religious faith that is lost with the onset of nihilism as Nietzsche conceives it, but also faith in reason (*Vernunft*) and science (*Wissenschaft*). For, as Nietzsche remarks in his November 1887–March 1888 notebook, unity, purpose, and truth are 'categories of reason'. These values are produced by and serve reason, and their devaluation is itself the result of reason being applied to them: 'The faith in the categories of reason is the cause of nihilism. We have measured the value of the world according to categories *that refer to a purely fictitious world*' (Nietzsche 1968: 13, 1999b: xiii. 49; Nietzsche's emphasis). With the advent of nihilism, '*every* belief, every considering-something-true' is lost – except, that is, for the belief that 'there simply is no *true world*' (Nietzsche 1968: 14, 1999b: xii. 354; Nietzsche's emphasis). The Nietzschean critique of nihilism is, then, not simply a critique of Christianity, but also a critique of reason, although in a sense distinct from, and more general than, Kant's. As Nietzsche puts it in his March 1887–November 1888 notebook, this devaluation of the highest values leads to nihilism as a sense of absolute valuelessness (*Werthlosigkeit*). With the abandonment of the (Christian) moral interpretation of the world, it is not simply one interpretation among others that is lost, but what was taken in the West to be the one true interpretation: 'One interpretation has collapsed; but because it was considered *the* interpretation it now seems as if there were no meaning at all in existence, as if everything were in vain' (Nietzsche 1968: 35, 1999b: xii. 212; Nietzsche's emphasis).

Nietzsche's theorization of the advent of nihilism thus conceived is beset by a number of obvious problems, which he would no doubt have sought to resolve had he ever written his projected 'History of European Nihilism'. First, in his notebooks he offers at least three distinct theses on the historical moment at which nihilism comes to prevail in the West. The first view is that nihilism commences at the moment when faith in the highest values is lost, when those values devaluate themselves. This devaluation Nietzsche presents as a nineteenth-century phenomenon, or, one might say, as modernity's experience of self-disillusionment. The 'Christian-moral' interpretation of the world 'leads to nihilism' because it turns back upon itself, judging itself to be untrue (Nietzsche 1968: 8, 1999b: xii. 125–6). This self-disillusionment may be seen as the most important of what Griffin terms the 'practical effects' of the French and industrial revolutions, with the 'myth of progress' having been undermined 'to a point where for many among its cultural elites modernity lost its utopian connotations and began to be constructed as a period of decline, decay, and loss' (Griffin 2007: 51). With the failure of the European revolutions of 1847–8 and a weakening of the Enlightenment sense of 'time pressing forward', of progress towards an ever better world, there emerges what Nietzsche terms 'Romantic pessimism', with 'its dark moods of nihilism and existential anguish' (Griffin 2007: 52). In fact, as a passage in *On the Genealogy of Morals* reveals, Nietzsche himself was far from clear about what precipitated this loss of faith in Enlightenment values. He proposes a range of explanations that reflect decidedly non-Enlightenment values, explanations that would prove fertile ground for Nazi ideologues in the 1920s and 1930s. Analysing what he terms a 'feeling of physiological inhibition', Nietzsche asserts that this feeling 'can have the most various origins'. He then offers the following possibilities:

> perhaps it may rise from the crossing of races too different from one another (or of classes – classes always also express differences of origin and race: European '*Weltschmerz*', the 'pessimism' of the nineteenth century, is essentially the result of an absurdly precipitate mixing of classes; or from an injudicious emigration – a race introduced into a climate for which its powers of adaptation are inadequate (the case of the Indians in India); or from the aftereffects of age and exhaustion in the race (Parisian pessimism from 1850 onward); or from an incorrect diet (the alcoholism of the Middle Ages; the absurdity of the *vegetarians* who, to be sure, can invoke the authority of Squire Christopher in Shakespeare; or from degeneration of the blood, malaria, syphilis, and the like (German depression after the Thirty Years' War, which infected half of Germany with vile diseases and thus prepared the ground for German servility, German pusillanimity). (Nietzsche 1989: 130–1; Nietzsche's emphasis)

There is much in Nietzsche's notebooks to suggest that he took nihilism to have arisen in the nineteenth century, as a development out of Romantic pessimism, and that he envisaged its exacerbation in the twentieth. However, by identifying Christianity as the 'most extreme form' of nihilism, Nietzsche not only locates nihilism's advent centuries earlier but he also implies that there were less extreme forms of nihilism prior to the nineteenth century. In their attempts to chart the history of nihilism as Nietzsche conceives it, both Martin Heidegger (in the 1930s and 1940s) and Gilles Deleuze (in the post-war era) take this view. According to Deleuze, the devaluation of the highest values experienced in the nineteenth century (that is, the crisis of modernity) is in fact a second phase in the unfolding of nihilism as conceived by Nietzsche. That phase is preceded, Deleuze argues, by an originary nihilism that lay in the very positing of those highest values, which occurred in Plato. This positing of the values of unity, purpose, and truth is nihilist in the sense that it negates existence as becoming (*Werden*); it is a devaluation whereby life (*Leben*) – a term the meaning of which is anything but clear in Nietzsche's thought – has 'the value of nil, the null value'. This originary nihilism is followed by a second phase – which Griffin describes as the crisis of modernity – when it is the 'nullity of values' as such that is experienced (Deleuze 1983: 148). In his 1940 lectures on 'European Nihilism', to which we shall turn in the next chapter, Heidegger anticipates Deleuze when he takes Nietzsche to chart a history of nihilism that commences with Platonism: 'Nihilism is not only the process of devaluing the highest values, nor simply the *withdrawal* of these values. The very positing of these values in the world is already nihilism' (Heidegger 1982b: 44; Heidegger's emphasis).

Just as one can find evidence in his notebooks to suggest that Nietzsche sees nihilism as a nineteenth-century phenomenon, so there is much in those same notebooks to support the view that he locates its advent in the origins of Western thought. He claims, for instance, that the Greek philosopher Pyrrho is 'more nihilistic' than Epicurus, and that the 'philosophers of Greece, e.g. Plato [...] represent one after the other the *typical* forms of decadence: the moral-religious idiosyncrasy, anarchism, nihilism (*adiaphora*), cynicism, obduracy, hedonism, reaction' (Nietzsche 1968: 239, 1999b: xiii. 277, 272; Nietzsche's emphasis). Indeed, he even goes so far as to assert that 'philosophers are always decadents – in the service of the *nihilistic* religions', since all philosophers prior to Nietzsche have posited a suprasensuous realm as the true world beyond the world of appearance as becoming (1968: 254, 1999b: xiii. 320; Nietzsche's emphasis). Furthermore, as mentioned above, Nietzsche often uses nihilism simply as a synonym for decadence and even for pessimism of the weak kind – that is, Schopenhauerian pessimism as opposed to Nietzsche's own 'pessimism of strength', which embraces the most fearsome thoughts about existence.

While Nietzsche never resolves the complications in his fragmentarily outlined 'history' of European nihilism, one can with some justification argue that he takes nihilism to reach its most extreme form when the nihilistic highest values devaluate themselves, which is to say when faith in the Enlightenment values of unity, purpose, truth, progress, reason, and science not only weakens but is lost altogether. In their place one finds a sense that, far from being an 'absolute value', the human being is merely an 'accidental occurrence in the flux of becoming and passing away' (1968: 9, 1999b: xii. 211).

If Nietzsche sometimes argues that the advent of nihilism lies in Plato, while at other times describing it as a nineteenth-century phenomenon, there are also a few occasions when he presents it as still to come. For instance, in an 1886–7 notebook entry he declares that 'The entire idealism of mankind hitherto is on the point of changing into nihilism – into the belief in absolute *worth*lessness, i.e., *meaning*lessness' (1968: 331, 1999b: xii. 313; Nietzsche's emphasis). And again, in his November 1887–March 1888 notebook, one finds the following passage:

> What I relate is the history of the next two centuries. I describe what is coming, what can no longer come differently: *the advent of nihilism*. This history can be related even now; for necessity itself is at work here. This future speaks even now in a hundred signs, this destiny announces itself everywhere [...]. (1968: 3, 1999b: xiii. 189; Nietzsche's emphasis)

And beyond these three, irreconcilable versions of the advent and history of nihilism to be found in Nietzsche's various notes on the subject over the last three years of his philosophical life (1886–8), there is a fourth, namely that nihilism has already been surpassed – by Nietzsche himself as 'the first perfect nihilist of Europe', the thinker who, as he puts it in his November 1887–March 1888 notebook, has 'lived through the whole of nihilism, to the end, leaving it behind, outside himself' (1968: 3, 1999b: xiii. 190).

A nihilism that commences with Plato, or with Christianity, or in the mid-nineteenth century, or that is still to come, or that has already been surpassed (if only by Nietzsche himself): to think these four histories together is not necessarily to fall into incoherence but to begin to grasp what Nietzsche means when, in his autumn 1885–autumn 1886 notebook, he declares that 'Nihilism stands at the door', and asks: 'whence comes to us this uncanniest [*unheimlichste*] of all guests?' (1968: 7, 1999b: xii. 125; translation modified). In part, this uncanniness (*Unheimlichkeit*) would lie in nihilism's history being characterized by precisely the kinds of complication outlined above, rendering any simple 'history of European nihilism' impossible. But that uncanniness is also to be found in Nietzsche's thinking of the appropriate response to nihilism – in short, how the required overcoming (*Überwindung*) of nihilism is to be accomplished.

In his autumn 1887 notebook, Nietzsche identifies two possible responses to the sense of worthlessness or meaninglessness experienced when the highest values devaluate themselves and the human being is rendered nothing more than an absurd mote in a universal process of becoming that is without purpose, truth, or unity. Both of these responses are themselves forms of nihilism, namely 'active' and 'passive' nihilism (see, for instance, Nietzsche 1968: 17–18, 1999b: xii. 350–1). While both of these forms might be seen as *reactive* in the sense that they are responses to an originary nihilism, Nietzsche always privileges active over passive nihilism, because he sees it as serving to clear the ground for the overcoming of nihilism through taking nihilism to its absolute limit, a limit that is not simply located beyond, but is of a fundamentally different kind to, the one reached by Christianity as the 'most extreme form' of nihilism. It is precisely on account of his championing of active nihilism that Nietzsche's later thought may be characterized as 'programmatic modernism' in Griffin's sense, which is to say a modernism that calls for renewal through a radical break with the past. In Nietzsche, this break takes the form of a 'revaluation of all values' (*Umwerthung aller Werthe*), and this revaluation depends on active nihilism as the negation of all existing values – in other words, the negation of the Enlightenment values underpinning modernity.

In an autumn 1887 notebook entry, Nietzsche offers the following general definition of the nihilist:

> A nihilist is a man who judges of the world as it is that it *ought* not to be, and of the world as it ought to be that it does not exist. According to this view, our existence (action, suffering, willing, feeling) has no meaning: the pathos of the 'in vain' is the nihilists' pathos – at the same time, as pathos, an inconsistency on the part of the nihilists. (1968: 318, 1999b: xii. 366; Nietzsche's emphasis)

And, in his November 1887–March 1888 notebook, he states that the essence of nihilism lies in the belief that 'all that happens is meaningless [*sinnlos*] and in vain; and that there ought not to be anything meaningless and in vain' (1968: 23, 1999b: xiii. 45). In fact, this clinging to the 'ought not' is characteristic of what Nietzsche elsewhere terms 'passive nihilism'; that is, a form of nihilism that cannot break free from the moral interpretation of phenomena. Passive nihilism cannot go 'beyond good and evil'; it is, Nietzsche argues, 'full of morality that is not overcome: existence as punishment, existence construed as error, error thus as a punishment – a moral valuation' (1968: 7, 1999b: xii. 126). In the face of this sense of existence as punishment or error, passive nihilism seeks (reasonably enough, it might seem) to reduce suffering to zero – this is the key trait shared by Buddhism, socialism, and Schopenhauer's philosophy of the will. In *The World as Will and Representation* (1819), a work whose influence on Nietzsche was profound, Schopenhauer argues for a denial (*Verneinung*) of the will in the interests of an escape from the suffering that it inflicts upon itself. That which lies beyond this suffering will is something for which 'we lack image, concept, and word' (Schopenhauer 1969: ii. 609). For the Nietzsche of the 1880s, such a negation of suffering is the negation of life itself.

Whereas passive nihilism remains attached to those highest values that are no longer sustainable, active nihilism is a 'violent force of *destruction*' directed against those values (Nietzsche 1968: 18, 1999b: xii. 351; Nietzsche's emphasis). Whereas passive nihilism is a 'decline and recession of the power of the spirit', active nihilism is 'a sign of increased power of the spirit' (1968: 17, 1999b: xii. 350). This power of the spirit is, however, not creative but rather unremittingly destructive. For this reason, it is what in his autumn 1887 notebook Nietzsche defines as 'genuine *nihilism*' (1968: 69, 1999b: xii. 468; Nietzsche's emphasis). As Michael Allen Gillespie puts it, active nihilism as Nietzsche conceives it 'is not content to be extinguished passively but wants to extinguish everything that is aimless and meaningless in a blind rage; it is a lust for destruction that purifies humanity' (Gillespie 1995: 179). This passion for destruction is clearly reminiscent of Bakunin and Nechaev's brands of Russian nihilism – or 'nihilism à la Petersburg', as Nietzsche describes it in the 1887 edition of *The Gay Science*: 'the *belief in unbelief* even to the point of martyrdom' (Nietzsche 1974: 289; Nietzsche's emphasis). Nietzsche argues, however, that active nihilism at its most extreme is in fact a form of affirmation (*Bejahung*). What it affirms is the thought of the 'eternal recurrence of the same' – which, in a notebook entry dated 10 June 1887 and headed 'European Nihilism', Nietzsche defines as 'existence as it is, without meaning or aim, yet recurring inevitably without any finale in nothingness'. The thought of eternal recurrence thus conceived is, according to Nietzsche, 'the most extreme form of nihilism: the nothing (the "meaningless"), eternally!' (Nietzsche 1968: 35–6, 1999b: xii. 213). The overcoming (*Überwindung*) of nihilism requires the affirmation of this most extreme form, since that is the affirmation of life as becoming (*Werden*), and thus of a new set of values. As the affirmation of eternal recurrence, active nihilism is the passage to the limit of nihilism.

Nihilism, then, as Nietzsche defines it, is radically ambiguous. Its history contains complications of the kind outlined above: it commences with Plato, with Christianity, in the mid-nineteenth century, is still to come, and has already been surpassed (if only by Nietzsche himself as the 'first perfect nihilist of Europe'). And it is at once the 'danger of dangers' – the destruction of all existing values (that is, the values underlying modernity) – and that through which this danger of dangers may be evaded: it is 'the sign of a crucial and most essential growth, of the transition to new conditions of existence' (1968: 69, 1999b: xii. 468). Through active nihilism, the ground can be cleared for a future governed by values that are not nihilistic in the sense that they do not constitute what in *On the Genealogy of Morals* Nietzsche terms a 'No to life', a negation of life conceived as a becoming that is not only without purpose, truth, or unity, but that takes the form of eternal recurrence (Nietzsche 1989: 19). The alternative to such active nihilism is an endless exacerbation of nihilism: 'Attempts to escape nihilism without revaluating our values so far […] produce the opposite, make the problem more acute' (Nietzsche 1968: 19, 1999b: xii. 476). Such a condition would be liminoid rather than liminal, an inescapable trap rather than a genuine threshold.

But how does the radical negativity of active nihilism lead to the overcoming of nihilism? How does unremitting negation become affirmation, and thereby evade the liminoid? How are old, nihilistic values revaluated into new ones? What are those new values, and to what can we appeal to underwrite them? According to Roger Griffin, Nietzsche theorizes active nihilism as 'endowed with a dialectical, self-overcoming component which is implicit in the description of Russian anarchists bent on overthrowing the Tsarist system as "nihilists". It now acquires a constructivist, futural thrust diametrically opposed to its connotations when it is used to refer to the absolute denial that life has any transcendent value or purpose' (Griffin 2007: 60). To be sure, Nietzsche's thinking of nihilism is no less futural than it is a sweeping characterization of Western culture from its origins to the late nineteenth century. But is Nietzsche's thinking of active nihilism dialectical? If the self-overcoming of nihilism requires the absolute negation of the moral interpretation, one cannot justify the claim that it is better for nihilism to be overcome than for it to continue – unless, that is, one has *already* accomplished the (non-moral) revaluation of all values. There would be no values to which active nihilism could appeal to justify its affirmation of eternal recurrence – unless that nihilism has *already* been overcome. It is for precisely this reason that Nietzsche finds it necessary to assert that he is the 'first perfect nihilist of Europe', the one who has 'even now lived through the whole of nihilism, to the end, leaving it behind, outside himself' (Nietzsche 1968: 3, 1999b: xiii. 190). Only from this position – 'outside himself' – can Nietzsche launch his critique of European nihilism without that critique being simply self-refuting. The very uncanniness (*Unheimlichkeit*) of nihilism to which Nietzsche himself alerts us lies in part in this outside being less a beyond-nihilism than its border. Indeed, rather than a *dialectical* self-overcoming of nihilism in which new values are produced through some kind of preserving negation – or, in Hegelian terminology, through an *Aufhebung* – of the old values, Nietzsche's thought in all its variability points towards a thinking of nihilism as that which remains forever 'at the door', neither inside nor outside, neither present nor absent, neither of the past nor of the future, neither threatening nor promising, but spectral, haunting, liminoid.

As for those new values that would enable the overcoming of nihilism, Ansell-Pearson rightly observes that, according to Nietzsche, 'the Greek experience of art can instruct us in how it is possible to overcome nihilism, not through a utopian politics or an eschatological religion, but through the cultivation of an affirmation of the tragic character of existence' (Ansell-Pearson 1994: 66). The privilege that Nietzsche accords to art as that which can lead the way in the overcoming of nihilism is anticipated by Schopenhauer, who sees the aesthetic as a means of escaping from the will, if only temporarily. According to Schopenhauer, the artist achieves a 'pure, true, and profound knowledge of the inner nature of the world', with this knowledge being for the artist an 'end in itself; at it he stops. Therefore it does not become for him a quieter of the will, as [...] in the case of the saint who has attained resignation; it does not deliver him from life for ever, but only for a few moments. For him it is not the way out of life, but only an occasional consolation [*einstweilen ein Trost*] in it, until his power, enhanced by this contemplation, finally becomes tired of the spectacle, and seizes the serious side of things' (Schopenhauer 1969: i. 267). In his first book, *The Birth of Tragedy* (1872), in which the influence of Schopenhauer is evident not least in his conception of art as offering metaphysical consolation (*Trost*), Nietzsche claims not only that art is 'the highest task and the true metaphysical activity of this life', but also, and on more than one occasion, that 'it is only as an *aesthetic phenomenon* that existence and the world are eternally *justified*' (Nietzsche 1967: 31–2, 52; Nietzsche's emphasis). In the 'Attempt at a Self-Criticism' that he wrote as a new preface for the 1886 edition of *The Birth of Tragedy*, Nietzsche claims – as he will again in *Ecce Homo* – that his first book should be understood in terms of an opposition more fundamental than the intra-aesthetic distinction between the Dionysian and the Apollonian, namely the distinction between art and morality, or between the aesthetic and the moral interpretation of the world:

> In truth, nothing could be more opposed to the purely aesthetic interpretation and justification of the world which are taught in this book than the Christian teaching, which is, and wants to be, *only* moral and which relegates art, *every*, to the realm of lies; with its absolute standards, beginning with the truthfulness of God, it negates, judges, and damns art. [...] Christianity was from the beginning, essentially and fundamentally, life's nausea and disgust with life, merely concealed behind, masked by, dressed up as, faith in 'another' or 'better' life. Hatred of 'the world', condemnation of the passions, fear of beauty and sensuality, a beyond invented the better to slander this life, at bottom a craving for the nothing [...]. (1967: 23; Nietzsche's emphasis)

As the most extreme form of moral interpretation, Christianity is for Nietzsche the most extreme form of passive nihilism: a 'craving for the nothing', a '*No* to life'. In direct opposition to this moral interpretation he places the aesthetic interpretation, which he takes to be the affirmation of life. The life that is affirmed in this aesthetic interpretation is, however, life as becoming (*Werden*), which is to say life characterized by 'semblance [*Schein*], art, deception, points of view, and the necessity of perspectives and error' (23). In short, the aesthetic interpretation interprets life *as art*, and in so doing produces new values, namely the values of appearance, fiction, deception, perspectivism, and error, values that are no longer those of modernity but arguably those of postmodernity, to which we shall turn in Chapter 5.

As the opposite of the moral interpretation of the world, the aesthetic interpretation is the opposite of nihilism. That this is Nietzsche's position becomes clear in an entry in his May–June 1888 notebook, in which he describes art as the 'only superior counterforce' (*einzig überlegene Gegenkraft*) to the nihilist devaluation of life. Art, he asserts in this fragment, is 'anti-Christian, anti-Buddhist, antinihilist *par excellence*' (Nietzsche 1968: 452, 1999b: xiii. 521). Art as the 'only superior counterforce' to nihilism is, however, art of a very particular kind, namely what in *The Birth of Tragedy* is theorized as the 'Dionysian'. The nihilism-countering force of this Dionysian art lies not in its power of imitation (mimesis) but rather in its power of transfiguration (*Verklärung*). In short, art for Nietzsche constitutes the overcoming of nihilism in its being free of morality and in its affirmation of life not through the faithful representation of that life but through its radical transformation.

A more detailed analysis of Nietzsche's conception of art as the counterforce to nihilism will be undertaken in Chapter 3, when we consider the relationship between aesthetic modernism and nihilism. As we shall see, Nietzsche's turn to art as the counterforce to nihilism in the various senses that the latter term has for him – the moral interpretation of phenomena, reason, science, the sense that life as becoming (*Werden*) lacks unity, meaning, purpose, or truth, an adherence to the Enlightenment values of modernity – recurs in many of the most important deployments of the concept of nihilism in the modernist critique of modernity, including both Heidegger's and Adorno's. Before considering the various ways in which art comes to be privileged as the counterforce to nihilism, and the implications of this privilege, however, it is first necessary to chart the influence of Nietzsche's conception of nihilism, and the role of that concept in the critique of modernity, within later philosophico-political discourses.

Perhaps the most telling element in the various discourses on nihilism in Nietzsche's wake is that, in each case, the *Unheimlichkeit* of nihilism – to which Nietzsche alerts us in the draft 'preface' on nihilism in his autumn 1885–autumn 1886 notebook – is either ignored or underestimated. As we shall see, each of the major deployments of the concept of nihilism in the first half of the twentieth century works on the assumption that nihilism is a concept that can be mastered, that it can be made to serve a critique of modernity. That this is not necessarily the case, that nihilism may be precisely that concept which cannot simply be co-opted in the interests of critique, while at the same time being the concept that lies at the very heart of the most thoroughgoing critiques of modernity, is the thought that will guide the analyses offered of both philosophical and aesthetic modernism in the following chapters. One of the principal spheres in which the uncanniness of nihilism manifests itself is the political, and it is no coincidence that in their engagement with the concept of nihilism, in their attempts to make it serve the cause of a critique of modernity, both philosophical and aesthetic modernism find themselves inextricably caught in forms of political extremism.

2

NIETZSCHE'S LONG SHADOW

Conservative revolutionaries in Germany: Spengler and Jünger

That the concept of nihilism came to play such a crucial role in the philosophico-political critique of modernity in the first half of the twentieth century in Germany is owing above all to the publication of Nietzsche's *The Will to Power* (1901; enlarged edition 1906), a volume that is not to be mistaken for the work that Nietzsche himself promised under that title in *On the Genealogy of Morals*. In the first part of *The Will to Power*, Nietzsche's editors collected together under the title 'European Nihilism' fragments from his notebooks relating to that topic, thereby giving the impression that the question of nihilism lay at the very heart of his later thought. The fundamental misconception to which this led is clear from the future Nazi philosopher Alfred Baeumler's postscript to the 1930 Kröner edition of *The Will to Power*, in which he declares it to be 'Nietzsche's philosophical *magnum opus*. All the fundamental results of his thinking are brought together in this book' (Baeumler in Nietzsche 1968: xiii). When Martin Heidegger came to lecture on Nietzsche in 1936–40, he paid far more attention to *The Will to Power* than to any work actually published by Nietzsche. Prior to Nietzsche's appropriation by Nazi ideologues in the 1930s, however, *The Will to Power*, and in particular the concept of nihilism therein, was already playing a determining role in the work of conservative revolutionaries in the immediate post-war period.

The interwar years saw the publication of a number of large-scale cultural critiques by intellectuals on the political Right in Germany, where the influence of Nietzsche was evident (albeit in ways that suggest highly questionable readings of his works), above all in the conception of modernity as nihilist and in the claim that the overcoming of this nihilism was possible only through a return to myth. Among the most influential of these critiques coming from what Peter Osborne has termed the 'conservative revolution' was Oswald Spengler's *The Decline of the West* (vol. 1, 1918; vol. 2, 1922). Matei Calinescu considers Spengler to be 'perhaps the most important philosopher of culture influenced by Nietzsche' (Calinescu 1987: 207–8), and, although Spengler declares that *The Decline of the West* is indebted to 'the philosophy of Goethe, which is practically unknown today, and also (but in a far less degree) to that of Nietzsche', he nonetheless identifies Nietzsche's call for a revaluation of all values as 'the most fundamental character of *every* civilization' (Spengler 1932: 49 n. 1, 351; Spengler's emphasis). Furthermore, Spengler claims that Nietzsche's announcement of the advent of nihilism is correct, and that 'Every one of the great Cultures knows it, for it is of deep necessity inherent in the finale of these mighty organisms. Socrates was a nihilist, and Buddha' (352). Of what he takes to be the three principal cultures – Buddhism, Stoicism, and socialism – Spengler states:

> What we have before us is the three forms of Nihilism, using the word in Nietzsche's sense. In each case, the ideals of yesterday, the religious and artistic and political forms that have grown up through the centuries, are undone; yet even in this last act, this self-repudiation, each several Culture employs the prime-symbol of its whole existence. The Faustian nihilist – Ibsen or Nietzsche, Marx or Wagner – shatters the ideals. The Apollonian – Epicurus or Antisthenes or Zeno – watches them crumble before his eyes. And the Indian withdraws from their presence into himself. (357)

Another highly influential work of the interwar years in which the concept of nihilism plays a key role is Ernst Jünger's *The Worker: Mastery and Form* (1932), in which Jünger develops an argument already articulated in broad outline in his essay *Total Mobilization* (1930). Jünger had come to prominence with the publication of *In Storms of Steel: From the Diary of a Storm Troop Leader* (1920), in which he recounts his experiences in the 1914–18 war. According to Jünger, this war was 'a historical event superior in significance to the French Revolution', since it marked the end of the bourgeois era and the dawn of a new age, which in *Total Mobilization* he terms the 'age of work' (*Arbeitszeitalter*). In this new age of 'total mobilization' (*totale Mobilmachung*), of 'masses and machines', 'the image of war as armed combat merges into the more extended image of a gigantic labor process' (Jünger 1993: 126). The distinction between soldier and worker, between a state of war and a state of peace, is dissolved, and everything is transfigured into energy (*Energie*) such that 'not a single atom is not in motion' (128). This energy is not creative but destructive, total mobilization conveying 'the extensively branched and densely veined power supply of modern life towards the great current of martial energy' (127). The soldier-worker supplants the bourgeois citizen – 'the modern armies of commerce and transport, foodstuffs, the manufacture of armaments – the army of labor in general' (126) – and all bourgeois values are transvaluated.

The first phase in this transvaluation is the annihilation of the bourgeois spirit. In the post-war era, according to Jünger, one sees 'the increasing curtailment of "individual liberty", a privilege that, to be sure, has always been questionable', and the disappearance of 'anything that is *not* a function of the state. We can predict a time when all countries with global aspirations must take up the process, in order to sustain the release of new forms of power' (127; Jünger's emphasis). The new values of the age of work are discipline, service, and above all the 'elemental', which is located in 'the primordial strength of the Volk' (*die Urkraft des Volkes*) and of which the bourgeoisie is not even aware (133). That the 'people' to which Jünger is referring here are first and foremost the German *Volk* becomes clear in *The Worker*, where the vision outlined in *Total Mobilization* is developed through a more extensive treatment of what he terms the form or figure (*Gestalt*) of the Worker (*Arbeiter*) as 'a new humanity' that is 'destined to mastery', a new humanity that marks a 'new dawn for Germany' (Jünger 1981: 72, 31; my translation). It is on the basis of this German model that the age of the Worker will become planetary.

The Worker opens with an unqualified attack on the bourgeoisie and its principal values: individualism, freedom, peace, the cult of reason, and, above all, security (see Jünger 1981: 54). According to Jünger, these Enlightenment values, established by the French Revolution, are now the values of a nihilism (*Nihilismus*) that is to be overcome through the assertion of new values, of which the principal ones are discipline and mastery as a form of service. Jünger insists that the *Gestalt* of the Worker is not a class, and has to be grasped metaphysically as a new form of humanity. While he refers on more than one occasion to the Worker as a race (*Rasse*), Jünger also insists that this concept is not to be understood biologically and that the mobilization effected by the Worker goes beyond all racial distinctions.

The influence of Nietzsche is evident in Jünger's claim that the Worker will exert mastery (*Herrschaft*) in order to overcome the nihilism of bourgeois liberalism, and also in the claim that a new will to power (*Machtwille*) is announcing itself. As Gerhard Loose observes, Jünger accepts without reservation Nietzsche's 'radical criticism, the total rejection of what this age stood for: the capitalist system (paramount importance attaching to material gain and social prestige), the democratic form of government (promoting, in his view, the rule of mediocrity), positivism and materialism (a shallow view of the world), and its art (lacking both substance and form)' (Loose 1974: 37). Indeed, it is likely that Jünger derived his core concept of the Worker (defined in contradistinction to the bourgeois individual) from two fragments (nos. 763 and 764), dating from 1887 and 1882, in Book Three of *The Will to Power*: 'Workers should learn to feel like soldiers' and 'The workers shall live one day as the bourgeois do now – but *above* them, distinguished by their freedom from wants, the *higher caste*: that is to say, poorer and simpler, but in possession of power' (Nietzsche 1968: 399; Nietzsche's emphasis).

Nietzsche's influence is apparent above all, however, in Jünger's reliance upon the concept of nihilism to characterize modernity, an era the origin of which he locates, unlike Nietzsche, in the French Revolution. While the terms *nihilistisch* and *Nihilismus* do not appear in the essay *Total Mobilization*, they are deployed repeatedly in *The Worker*: with regard to the idea of progress, which Jünger rejects as a bourgeois value (see Jünger 1981: 50); the concept of Romantic protest as a flight from the real (60); bourgeois values in general (99); the cult of hygiene, sun worship, and sport (110); a general process of decomposition (132); the essence of technology (172); and those oppositions that give rise to endless dialectical discussion (242).

The supplanting of the bourgeoisie with the total mobilization of the Worker may be seen – in Nietzschean terms – as an active nihilism working against the reactive nihilism of an age in which mobilization is always limited, this limitation being, in Jünger's opinion, the principal reason for Germany's defeat in the 1914–18 war. As an active nihilism, technologization as such affects the totality of relations and no bourgeois value can resist it. What Jünger refers to as the 'destructive phase' (173) in the total mobilization of the Worker is that phase through which it is necessary to pass in order to overcome the nihilism of modernity and its values as established by the French Revolution. In this 'destructive phase', technology (*Technik*) as active nihilism produces anarchy, undoing all existing relations, destroying all faith and being resolutely anti-Christian; it is, Jünger claims, 'the manner and the means through which the *Gestalt* of the Worker mobilizes the world' (160). The accomplishment of this overcoming of nihilism will result, according to Jünger, in a new order (*Ordnung*) in which 'that race achieves mastery which knows how to speak the new language, not in the sense of mere understanding, progress, utility, comfort, but as an elemental language [*Elementarsprache*]' (173). As Loose observes, in contrast to Hitler's *Mein Kampf* (1925–6) and Alfred Rosenberg's *Myth of the Twentieth Century* (1930), *The Worker* at least 'presents the case of militant totalitarianism, or of aggressive nihilism, clearly and honestly' (Loose 1974: 38). To that should be added, however, that, unlike both Hitler and Rosenberg, Jünger does not present 'the Jew' as the principal embodiment of the decadence of modernity.

Jünger emphasizes that the annihilation of bourgeois society, and its replacement by a Worker state that draws its power from the elemental, is not to be mistaken for a purely political or economic event; rather, it is strictly metaphysical in nature. In short, the rise of the Worker is not the rise of a new class, but a revolution in Being (*seinsmäßige Revolution*) (Jünger 1981: 112). In his theorization of the Worker as a new form of humanity, Jünger insists upon the annihilation of the classical conception of the nation-state. His argument is nonetheless a form of philosophico-political modernism, and, crucially, one that remains tied to the concept of the state. The *Gestalt* of the Worker is located first and foremost in Germany, since, as Jünger announces at the beginning of *The Worker*, the German was never a good bourgeois citizen; it is the 'destiny' (*Schicksal*) of Germany that depends upon the emergence of this new *Gestalt*, and this destiny promises a 'new dawn for Germany' (31). With that new dawn, and the 'heroic realism' that arises with it, the overcoming of nihilism commences. The value of values for this new age is a mastery (*Herrschaft*) rooted in the elemental, a mastery exercised not by individuals but by a type thought in terms of the German *Volk*.

Although he deploys the term 'nihilism' at key moments in *The Worker*, Jünger's most developed analysis of the concept is to be found in his post-war work, *Across the Line* (1950), in which his position is strikingly different from the one articulated in *The Worker*, doubtless in no small part on account of what had occurred between 1933 and 1945. That this is the case is evidenced by the fact that, whereas it is bourgeois liberalism that is identified as nihilist in *The Worker*, in a diary entry of 16 October 1943 Jünger refers to Reinhard Heydrich as the 'nihilist-in-chief' responsible for the killing of the Jews in the Lodz ghetto. Furthermore, in his wartime diary, Jünger records an encounter with the French writer Louis-Ferdinand Céline, whose extreme anti-Semitism led Jünger to see in him an incarnation of 'the monstrous powers of nihilism', and to write: 'Céline with his dirty fingernails – I'm now entering a phase in which the sight of nihilism is becoming physically intolerable' (Jünger quoted in Nevin 1997: 186).

The two points of departure for Jünger's thinking of nihilism in *Across the Line* are Nietzsche and Dostoevsky, and from both he takes a determination of nihilism as a loss of values and a theory of its overcoming – a crossing of the line that marks the 'zero point' of nihilism; that is, its most extreme form. Jünger identifies Nietzsche and Dostoevsky as the two great sources of knowledge on nihilism because they teach us to think it as a phase in a 'spiritual process' through which it is necessary to pass (see Jünger 1980: 239; my translation). This phase cannot be avoided, but it can be surpassed. Although the term 'nihilism' has, according to Jünger, almost always been deployed polemically, nihilism properly understood is in fact a 'great destiny', a 'fundamental power', albeit one that constitutes the greatest threat to the human (244). It has been *the* 'great theme' for the past century, and finds expression in the works of, among others, Paul Verlaine, Marcel Proust, Georg Trakl, Rainer Maria Rilke, Lautréamont, Nietzsche, Arthur Rimbaud, Maurice Barrès, Joseph Conrad, William Faulkner, André Malraux, Georges Bernanos, Ernest Hemingway, Franz Kafka, Oswald Spengler, Gottfried Benn, Henry de Montherlant, and Graham Greene. Crucially, in the post-war era it is no longer in political modernism but rather in *aesthetic* modernism that Jünger finds an authentic confrontation with nihilism: 'Common to all of them [the above writers] is the experimental, the provisional attitude, and the knowledge of the dangerous situation, the great threat' (253). Nihilism, he insists, is to be distinguished from sickness, evil, and chaos. Relating what he presents as a diagnosis of nihilism to his earlier treatise on the Worker as the

Gestalt of contemporary man, Jünger argues that, far from being a form of chaos, nihilism can harmonize with 'vast worlds of order' (249), and that active nihilism ultimately produces human beings who function like 'iron machines' – as the pure technologization of the human – indifferent to the catastrophe that is modern European history (252).

Taking up Nietzsche's identification of nihilism as the 'uncanniest of all guests', Jünger applies this to contemporary historical events, arguing that the unfolding of nihilism has now disclosed its 'uncanniest features' in the death camps (250). That said, Jünger believes that the step beyond the zero point of absolute nihilism has already been taken. Like Heidegger, Jünger identifies the poet (*Dichter*) and the thinker (*Denker*) as those figures who possess the spiritual force to surpass nihilism. The three 'fundamental powers' through which this overcoming of nihilism can be accomplished are the wilderness (but not a *romantic* wilderness), eros (which is not to be confused with a purely technologized sex of the kind to be found in the works of Henry Miller), and art (the authenticity of which lies in its self-consciousness, its rationality, and its critical self-control). Both the thinker and the poet risk themselves 'on the border of nothingness'; both endure 'the immense power of nothingness'; and both engage in a 'sovereign struggle' with this prevailing nothingness (277–9). For Jünger, it is only through this struggle (*Kampf*) that new values can come into being. Like Nietzsche before him, then, the Jünger of the post-Second World War, post-Holocaust era comes to see art as a form (arguably the most genuine form) of resistance to the prevailing nihilism of modernity.

The struggle against nihilism, or from Nazism to poetry: Heidegger

Of all the figures in the history of the deployment of the concept of nihilism in the critique of modernity, it is the philosopher Martin Heidegger (1889–1976) who makes by far the greatest use of the term – far more, in fact, than Nietzsche, the critique of whose thought comes to occupy a central place in Heidegger's thinking in the 1930s. Furthermore, Heidegger commits himself unreservedly to the conception of modernity as nihilist in its very essence. The timing of Heidegger's first deployment of the term *Nihilismus* is anything but coincidental. Indeed, it is decidedly philosophico-political in nature.

Heidegger presents his own engagement with Nietzsche's thought in the 1930s and early 1940s as a counter-movement to the National Socialist appropriation of that thought, an appropriation evident in, for instance, the work of Alfred Baeumler, professor of philosophy in Berlin from 1933 to 1945 and author of *Nietzsche the Philosopher and Politician* (1931). Heidegger's engagement with Nietzsche in four lecture series delivered between 1936 and 1940 at Freiburg University, and also in a series of essays and longer treatises, concentrates on the fragments collected in *The Will to Power* and in particular those on the concept of nihilism. In addition to the lectures, which were first published over two decades after their delivery, in the two-volume *Nietzsche* in 1961, the essays and treatises in which Heidegger articulates his own conception of nihilism include 'The Word of Nietzsche: "God Is Dead"' (written in 1943), 'Nihilism as Determined by the History of Being' (written in 1944–6), and 'The Essence of Nihilism' (written in 1946–8).

In a letter of 4 November 1945 to the rector of Freiburg University, Heidegger defends himself against the charge of his having been an active supporter of Nazism. He even goes so far as to claim that in his lectures on Nietzsche he engaged in an act of 'spiritual resistance' against National Socialism, and that he did so precisely through a critical engagement with Nietzsche's thinking on nihilism. According to the Heidegger of late 1945, National Socialism is itself a manifestation of nihilism:

> Beginning in 1936 I embarked on a series of courses and lectures on Nietzsche, which lasted until 1945 and which represented in even clearer fashion a declaration of spiritual resistance. In truth, it is unjust to assimilate Nietzsche to National Socialism, an assimilation which – apart from what is essential – ignores his hostility to anti-Semitism and his positive attitude to Russia. But on a higher plane, the debate with Nietzsche's metaphysics is a debate with *nihilism* as it manifests itself with increased clarity under the political form of fascism. (Heidegger in Wolin 1993: 65; Heidegger's emphasis)

While it is certainly the case that the lectures on Nietzsche to which Heidegger refers in this letter do counter the biologist and racialist interpretation of Nietzsche being promulgated by National Socialist ideologues in the 1930s, these lectures are neither

Heidegger's first public engagement with Nietzsche's thought, nor the first occasion on which he deploys the term 'nihilism'. And, at the time of that first deployment, Heidegger presents National Socialism not as a clearer manifestation of nihilism, but precisely as the privileged form of resistance to it. By accepting the rectorship of Freiburg University shortly after the Nazi Party came to power in Germany, Heidegger committed himself both to the leader principle (*Führerprinzip*) and to the political alignment of the university with the Nazi Party (the so-called *Gleichschaltung*). In a speech delivered in February 1934, Heidegger describes Hitler as a man of 'unprecedented resolve', whose 'glorious will' would ensure that 'the German people may once again find its growing unity, its true worth and true power, and may procure thereby its endurance and greatness as a work State' (Heidegger 2003b: 15). Neither in his 27 May 1933 Rectoral Address, *The Self-Assertion of the German University*, nor in the various political speeches delivered during his rectorship of 1933–4, however, does Heidegger make use of the term 'nihilism'. Indeed, despite its being a statement of support for National Socialism, Heidegger's Rectoral Address was in fact denounced in the *Neue Zürcher Zeitung* as 'the expression of an abysmal and destructive nihilism' (quoted in Farías 1989: 110). And, in an article published in the National Socialist periodical *Volk im Werden* in 1934, Ernst Krieck, who at the time was professor of philosophy and education studies at the Pedagogical Academy in Frankfurt, claimed that 'The meaning of [Heidegger's] philosophy is downright atheism and metaphysical nihilism of the kind that used to be represented in our country mainly by Jewish literati – in other words, an enzyme of decomposition and dissolution for the German people' (quoted in Safranski 1998: 268).

In defending himself against these charges of nihilism, the second of which places him in the same highly dangerous category as 'Jewish literati', Heidegger takes up the term in his 1935 lecture course, *Introduction to Metaphysics* (first published in 1953). He does so in relation to the possibility of a thinking and speaking of the nothing (*das Nichts*) already broached in *What Is Metaphysics?* (2009), but now addressed by way of the concept of nihilism. Looking back in 1955 on the reception of *What Is Metaphysics?*, Heidegger observes that 'people have seized upon and extracted "the" nothing and made the lecture into a testament to nihilism' (Heidegger 1998: 317). In *What Is Metaphysics?* itself, Heidegger declares that, for science (*Wissenschaft*), the nothing is 'an outrage and a phantasm' (84). Six years later, however, after certain unnamed 'people' have turned his 1929 Inaugural Lecture into a 'testament of nihilism', Heidegger claims that, from the perspective of 'logical' or 'scientific' thinking, to speak about the nothing is 'pure nihilism' (Heidegger 2000b: 26). According to Heidegger, that charge is not only misplaced, but itself comes from what he describes as 'real nihilism' – an expression that will recur in the 1944–6 treatise 'Nihilism as Determined by the History of Being'. In his 1935 lectures, then, Heidegger seeks to defend himself by turning the highly dangerous charge of nihilism back against those who have directed it at him. Neither in *Introduction to Metaphysics* nor in the many later texts in which he explicitly addresses the question of nihilism, however, does the logic governing this primal scene of charge and counter-charge become the object of Heidegger's own analysis. This alone is enough to show that he fails to take adequate account of nihilism in what Nietzsche terms its 'uncanniness'.

Heidegger's deployment of the term in the 1935 lecture course is the point of departure for a complete redetermination of the concept of nihilism inherited from Nietzsche. This redetermination involves the turning back of the charge of nihilism not only against those who would accuse Heidegger himself of nihilism but also against Nietzsche's own determination of nihilism as devaluation (*Entwerthung*). To ask after the nothing is, Heidegger argues, the very opposite of nihilism, since it is, albeit indirectly, to ask after Being (*Sein*) – and that, according to Heidegger, is the sole path towards the overcoming (*Überwindung*) of 'real nihilism', the essence of which lies in the forgetting of Being (*Vergessenheit des Seins*). It is not in the attempt to think the nothing that this 'real nihilism' is at work, but rather:

> where one clings to current beings and believes it is enough to take beings, as before, just as the beings that they are. But with this, one rejects the question of Being and treats Being as a nothing (*nihil*), which in a certain way it even 'is', insofar as it essentially unfolds. Merely to chase after beings in the midst of the oblivion of Being – that is nihilism. Nihilism thus understood is the ground for the nihilism that Nietzsche exposed in the first book of *The Will to Power*. (217)

Leaving aside the fact that, strictly speaking, Nietzsche exposed nothing about nihilism in the 'first book' of *The Will to Power*, since that book consists of a collection of notes, drafts, and fragments drawn together by the work's editors (on the basis of one among many outlines for such a work) to give the impression that Nietzsche had undertaken such an exposure, the above statement encapsulates Heidegger's own countering redetermination of nihilism. For Heidegger, the essence of nihilism lies not in the devaluation of the highest values, as Nietzsche claims, but rather in the negation of the difference between Being and beings, a negation that Heidegger finds throughout Western thought from as early as Plato. Moreover, far from constituting the overcoming of nihilism, Nietzsche's own thinking of Being in terms of value is, according to Heidegger, itself the consummation of nihilism.

At stake for Heidegger at the time he first deploys the concept of nihilism publicly is the future of what he terms 'Europe' and the 'West'. If Europe is 'always on the point of cutting its own throat', in 1935 this threat from within is conjoined with a threat from far beyond Germany's borders, since Europe as a whole is caught between the 'great pincers' of Russia and America. Seen 'metaphysically', these two threats – communism and capitalism – are one and the same: 'the same hopeless frenzy of unchained technology and of the rootless organization of the average man' (40). As he reiterates thirty years later in his 1966 interview with *Der Spiegel*, Heidegger sees Russia and America as forces of nihilism in the form of an 'unchained technology' that 'tears men loose from the earth and uproots them' (Heidegger 2003b: 37). In neither Russian communism nor American capitalism does one find a 'genuine confrontation' with technology as that which uproots and renders the human Dasein homeless (*heimatlos*). As the forgetting of Being, then, nihilism has to be thought as a subjection to technology as that which deracinates, detaches the human being from its proper *Heimat*. To see this deracination not as a liberation but as the decline that it is – this, according to Heidegger, requires 'spiritual strength'.

When Heidegger first publicly deploys the term 'nihilism', then, it stands for a technological deracination that has already spread 'throughout the whole earth' and that finds its most extreme forms in 'Russia' and 'America'. To counter this nihilism requires a 'genuine confrontation' with technology. The responsibility for this confrontation falls not to human beings in general, not even to Europeans, but rather to a particular people (*Volk*), namely the people of the 'centre', the 'metaphysical people', the 'historical people', the 'most endangered people' – in short, the German people (Heidegger 2000b: 41). To counter nihilism is, according to Heidegger, that people's vocation (*Bestimmung*). What is required, if Europe is to avoid complete annihilation, is 'the development of new, historically *spiritual* forces' in that people (41; Heidegger's emphasis). In his privileging of the German people in the struggle against nihilism, Heidegger is anticipated not only by Jünger's *The Worker* (1932), but also by two works of the late nineteenth century: Paul Lagarde's *German Writings* (1878) and Julius Langbehn's *Rembrandt as Educator* (1890). For the Heidegger of 1935, however, that 'people' is not simply German but National Socialist, the 'inner truth and greatness' of this movement lying, according to Heidegger, precisely in its encounter with technology – in other words, in its confrontation with 'real nihilism'. Four decades later, Heidegger is prepared to claim that 'with Marx the position of the most extreme nihilism is reached' (Heidegger 2003a: 77). His insistence that such an interpretation is 'not political', but concerned with 'being and the manner in which it destines itself', fails to take account of the fact that nihilism is a concept that is never purely philosophical, but is always philosophico-political. No one's theory of nihilism reveals this more clearly than does Heidegger's.

In his 1935 lecture series *Introduction to Metaphysics*, Heidegger claims that National Socialism as the movement which genuinely confronts nihilism is to be distinguished from anything championed by those who currently 'peddle' its 'philosophy' (Heidegger 2000b: 213). Again, in his summer 1936 lecture course on Schelling, Heidegger identifies Hitler, together with Mussolini, as 'The two men who in different ways introduced a countercurrent to nihilism', and who both 'learned, in essentially different ways, from Nietzsche'. However, he also places both Hitler and Mussolini among those for whom 'the authentic realm of Nietzsche's metaphysics still did not come into play' (quoted in Pöggeler 1993: 220). It is this 'authentic realm' that Heidegger sets out to chart in his own lectures and treatises of the later 1930s and 1940s.

In the first of his Nietzsche lecture courses, *The Will to Power as Art* (1936), Heidegger identifies Nietzsche as 'the first to recognize and proclaim with full clarity' the occurrence of nihilism (Heidegger 1979: 90). As we have seen, however, in *Introduction to Metaphysics* Nietzsche's thinking of Being in terms of value is presented as the consummation of the forgetting of Being. And, in the section entitled 'Nihilism' in *Contributions to Philosophy* (written in 1936–8), Heidegger declares that Nietzsche's interpretation of nihilism as devaluation is 'provisional, in spite of its importance'. The task for thought now is to grasp nihilism 'more fundamentally as the essential consequence of the abandonment of being' (Heidegger 1999: 96). The difficulty of this task is owing not least to the fact that it is precisely when nihilism is taken to have been overcome that it is at its most extreme. Thought in Nietzschean terms, as the loss of all goals:

> the greatest nihilism is precisely where one believes to have goals again, to be 'happy', to attend to making equally available the 'cultural values' (movies and seaside resort vacations) to all the 'people' – in the drunken stupor of 'lived-experience' – precisely there is the greatest nihilism: methodically disregarding human goallessness, being always ready to avoid every goal-setting decision, anxiety in the face of every domain of decision and its opening. (97)

The future direction of Heidegger's thinking of nihilism is indicated here by his insistence upon the danger of false overcomings, and by the emphasis upon Being as that which has abandoned beings, rather than upon those beings' forgetting of Being: 'Be-ing has so thoroughly abandoned beings and submitted them to machination and "lived-experience" that those illusive attempts at rescuing Western culture and all "culture-oriented politics" must necessarily become the most insidious and thus the highest form of nihilism' (97–8). Whereas in *Introduction to Metaphysics* the first step in the overcoming of nihilism is identified as a going 'to the limit of Nothing', by the time of *Contributions to Philosophy* 'The preparation for the overcoming of nihilism begins with the fundamental experience that man as founder of Da-Sein is *used* by the godhood of the other god' (98; Heidegger's emphasis). The emphasis has shifted, then, away from an active countering of nihilism of the kind Heidegger had originally found in National Socialism to a vision of man as giving himself over to the 'other god', a god that remains to come and that appears irreducible to any political movement or leader.

Heidegger's first sustained attempt to redetermine the concept of nihilism against Nietzsche occurs in his 1940 lecture course, *European Nihilism*. His position here is fundamentally the same as in 1935: Nietzsche's conception of nihilism as devaluation is itself nihilistic, and Nietzsche's proposed overcoming of nihilism through a revaluation of all values is in fact the consummation of nihilism as the forgetting of Being: 'Nietzsche knew and experienced nihilism because he himself thought nihilistically. Nietzsche's concept of nihilism is itself nihilistic' (Heidegger 1982b: 22). We have seen in the previous chapter that Nietzsche's own fragments relating to a 'History of European Nihilism' offer a complicated vision of nihilism as that which began with Plato but also in the nineteenth century, and as that which is already over but also still to come. Heidegger's own history of nihilism is far clearer: as the forgetting of Being, nihilism originates in Plato, takes a new form with Descartes (where Being is redetermined as *subiectum*), and reaches its consummation in Nietzsche (where Being is redetermined as will to power). That said, Heidegger also claims that 'for a comprehension of the essence of nihilism there is little to be gained by recounting the history of nihilism in different centuries and depicting it in its various forms', since nihilism does not 'have' a history. Rather, 'Nihilism *is* history' (53; Heidegger's emphasis).

For all his attempts to free the thinking of nihilism from any such history, however, contemporary history arguably impacts upon Heidegger at various key moments. In the wake of the German defeat at Stalingrad in February 1943, for instance, he shifts from describing nihilism as the 'forgetting' or 'abandonment' of Being, to seeing it in the treatise 'The Word of Nietzsche' as the 'radical killing' of Being, with Nietzsche's interpretation of nihilism as devaluation being 'murderous in a most extreme sense' (Heidegger 1977: 108). Given the ever-worsening military position of the Germans as the 'metaphysical people' of the 'centre', this rhetorical shift is surely no coincidence:

> Value-positing has brought down and slain beneath itself – and has therefore killed as that which *is* for itself – all that *is* in itself. This ultimate blow in the killing of God is perpetrated by metaphysics, which, as the metaphysics of the will to power, accomplishes thinking in the sense of value-thinking. However, Nietzsche himself no longer recognizes this ultimate blow, through which Being is struck down to a mere value, for what the blow is when thought with respect to Being itself. (107–8; Heidegger's emphasis)

The 1943 treatise 'The Word of Nietzsche' marks a culmination in Heidegger's thinking of nihilism precisely on account of this refiguration of nihilism as radical killing. In the three major texts devoted explicitly to nihilism after 1943 – 'Nihilism as Determined by the History of Being' (written in 1944–6), 'The Essence of Nihilism' (written in 1946–8), and 'On the Question of Being' (1955) – he moves towards figurations of nihilism in which the emphasis falls much more heavily upon the history of Being itself as that which determines nihilism. This counters any sense of a call to the *active* resistance of nihilism. In short, Heidegger moves from a programmatic to an epiphanic form of philosophico-political modernism.

By the time of the 1944–6 treatise 'Nihilism as Determined by the History of Being', Heidegger has abandoned the claim that either the National Socialist movement or even the German *Volk* can effectively counter nihilism: 'The essence of nihilism is not at all the affair of man, but a matter of Being itself, and thereby of course also a matter of the *essence* of man, and only in *that* sequence at the same time a human concern' (Heidegger 1982b: 221; Heidegger's emphasis). Since it is Being itself that 'has brought [nihilism] to pass', any attempt to overcome nihilism through struggle (*Kampf*) in the age of its consummation will necessarily prove futile:

> The struggle [*Kampf*] over nihilism, for it and against it, is engaged on a field staked out by the predominance of the nonessence of nihilism. Nothing will be decided by this struggle. [...] The will to overcome nihilism mistakes itself because it bars itself from the revelation of the *essence* of nihilism as the history of the default of Being, bars itself without being able to recognize its own deed. (239–40; Heidegger's emphasis)

In his 1935 lecture series, *Introduction to Metaphysics*, Heidegger identifies the symptoms of nihilism as the 'spiritual decline of the earth', the 'darkening of the world', the 'flight of the gods, the destruction of the earth, the reduction of human beings to a mass, the hatred and mistrust of everything creative and free', 'a *disempowering of the spirit*, its dissolution, diminution, suppression, and misinterpretation', and 'the onslaught of what we call the demonic' (Heidegger 2000b: 40, 47, 49; Heidegger's emphasis). A decade later, in the 1944–6 treatise, he takes the situation to be, if anything, even worse. The demise of the Third Reich, of which he makes no mention, appears to have marked no improvement at all, in Heidegger's eyes. The age of nihilism is the 'age of confusion, of violence and despair in human culture, of disruption and impotence of willing. Both openly and tacitly, boundless suffering and measureless sorrow proclaim the condition of our world a needful one' (Heidegger 1982b: 245). It is the age of the 'actual an-nihil-ation [*Ver-nichts-ung*] of *all* beings, whose violence, encroaching on all sides, makes almost every act of resistance futile' (229). The homelessness (*Heimatlosigkeit*) brought about by nihilism is no longer the homelessness of one 'historical people' – the German *Volk* – but 'the homelessness of historical man within beings as a whole. The "where" of a dwelling in the midst of beings as such seems obliterated, because Being itself, as the essential occurring of every abode, fails to appear' (248).

Like Nietzsche before him, however, Heidegger sees nihilism as profoundly ambiguous. Whereas Nietzsche distinguishes between active and passive nihilism, Heidegger thinks nihilism not only as 'an-nihilation' but also as a secret, an enigma, and a promise (226–8). We have seen in the previous chapter that, in order to justify a critique of nihilism as devaluation, Nietzsche has to appeal to 'new values'. Heidegger's critique of nihilism as the forgetting of Being (and the *Heimatlosigkeit* produced by this forgetting) cannot be justified by an appeal to values, since, according to Heidegger, evaluative thinking is itself nihilism. Rather, one has to appeal to what at the end of the 1944–6 treatise he terms the 'worth' (*Würde*) of Being, which 'does not consist in being a value, even the supreme value' (Heidegger 1982b: 250). Whether Heidegger manages to distinguish clearly between worth and value remains no less questionable, however, than the success of his distinction between a metaphysical and a political thinking of nihilism.

Heidegger's last major engagement with the concept of nihilism occurs in 1955, in an essay written in response to Ernst Jünger's *Across the Line* (1950). As we have seen, Jünger diagnoses nihilism in terms of the total mobilization of technology on a planetary scale. While agreeing with Jünger on the relation between technology and nihilism, Heidegger nonetheless sees Jünger's thinking of nihilism as being, like Nietzsche's (on which it draws so heavily), itself nihilist because it remains a thinking of nihilism in terms of values. Furthermore, Heidegger takes issue with what he considers to be Jünger's Nietzschean thinking of the overcoming (*Überwindung*) of nihilism, arguing instead for what he terms its *Verwindung*. Variously translated as 'recovery', 'surpassing', 'incorporation', 'healing', and even 'distortion', *Verwindung* is to be understood not as a movement across that line marking the borders of nihilism's terrain, but rather as a turning back towards the essence of nihilism. As Heidegger puts it: 'Turning in into its essence is the first step through which we may leave nihilism behind us. The path of this turning takes the form and direction of a turning back' (Heidegger 1998: 319).

For all the differences that Heidegger seeks to establish between his own thinking of nihilism and those of Jünger and Nietzsche before him, there remain a number of key similarities between them. First and foremost, Heidegger follows Jünger in thinking of modernity as nihilist in its very essence, with technology being that which radically transforms the essence of the human being. In the 1930s, both Jünger and Heidegger locate the counterforce to this all-pervasive nihilism in forces that are in large part political in nature. According to Jürgen Habermas, Heidegger follows Hegel in seeing art as having reached its end in romanticism, and is thus faced 'first of all with the task of putting philosophy in the place that art occupies in Nietzsche (as a countermovement to nihilism). [...] Nietzsche had entrusted the overcoming of nihilism to the aesthetically revived Dionysian myth. Heidegger projects this Dionysian happening onto the screen of a critique of metaphysics, which thereby takes on world-historical significance' (Habermas 1987: 97–9). In the course of the catastrophic political events of the 1930s and early 1940s, however, Heidegger, like Jünger, in fact comes to seek what Nietzsche terms the 'superior counterforce' to nihilism not in politics and, arguably, not even principally in philosophy, but, like Nietzsche before him, in a particular form of the aesthetic. For, from the mid-1930s onwards, Heidegger privileges poetry (*Dichtung*) as the form of language that recalls and names Being. *Dichtung* as Heidegger conceives it is not to be confused with 'literature'. As he makes clear in his 1966 interview with *Der Spiegel*, Heidegger considers 'present-day literature' the very antithesis of *Dichtung*, the former being a manifestation of nihilism precisely as a will to *homelessness*: 'According to our human experience and history, at least as far as I see it, I know that everything essential and everything great originated from the fact that man had a home and was rooted in a tradition. Present-day literature for example is predominantly destructive' (Heidegger 2003b: 37).

Heidegger's turn to poetry as a privileged form of resistance to nihilism occurs in the mid-1930s, when he lectures not only on Nietzsche but also on the German poet Friedrich Hölderlin (1770–1843). It is above all in Hölderlin's poetry that Heidegger finds the counterforce to nihilism as the forgetting of Being, since, according to Heidegger, that poetry recalls Being by naming it as 'the holy' (*das Heilige*). In other words, whereas Nietzsche champions as the counterforce to nihilism a Dionysian art that, after his disillusionment with Richard Wagner, is not tied to any one writer or artist beyond himself, Heidegger champions the work of a single poet. Hölderlin is, for Heidegger, the true poet of Being – and of the German people in their Germanness. In each of the lecture courses on Hölderlin delivered between 1935 and 1942, Heidegger insists that true poetry is not patriotic (*vaterländisch*) or political (*politisch*) as these concepts are commonly understood. In the 1934–5 lectures, he declares that '*The "fatherland" is Being itself*' (Heidegger 1989: 121–2; Heidegger's emphasis; my translation), and, in the 1941–2 lectures, that the fatherland which is founded by poetry is not to be mistaken for a 'present political constellation' (Heidegger 1982a: 47; my translation). And yet, Heidegger's public turn to *Dichtung* – and, more precisely, to Hölderlin's poems 'Germania' and 'The Rhine' – in the winter semester of 1934–5 is certainly not simply a flight from the political. Indeed, the abiding concerns of the Hölderlin lectures between 1934 and 1942 may be seen as political through and through, and thus as fully justifying Philippe Lacoue-Labarthe's claim that Heidegger's '"Hölderlinian" preaching is the continuation and prolongation of the philosophico-political discourse of 1933' and the production of what Lacoue-Labarthe terms a 'national aestheticism' (Lacoue-Labarthe 1990: 12, 53).

For all his insistence upon the difference between his own thinking of nihilism and that of Nietzsche, then, Heidegger follows Nietzsche in at least three key respects. First, he conceives of modernity, which for Heidegger extends at least as far back as the Cartesian revolution, as essentially – and, indeed, increasingly – nihilist. Secondly, he seeks to redetermine nihilism, turning it against the very discourse from which he appropriates the term. And, thirdly, he comes to locate the counterforce to nihilism in a particular form of art. In the next chapter, we shall return to Heidegger's conception of poetry (*Dichtung*) as the counterforce to nihilism in order to consider it within the context of a more general movement of thought that takes aesthetic modernism to be such a counterforce. Before doing so, however, it is first necessary to consider other major post-Second World War philosophical deployments of the concept of nihilism in the critique of modernity that, while indebted in different ways to Nietzsche's thinking of nihilism, come not only from the political Right, as do Jünger's and Heidegger's, but also from the Left.

National Socialism as nihilism: Adorno

We have seen that when, in 1935, he first uses the term 'nihilism' in a public discourse, Heidegger identifies National Socialism as that movement which has the 'inner truth and greatness' to counter the nihilism of modernity, and that this countering would take the form of a confrontation with technology. In part, this position was clearly in line with the Nazi ideologues' own view of the relation between National Socialism and nihilism, even if they did not conceive of the latter in terms of technology. For Heidegger, nihilism manifested itself in 'Russia' and 'America', communism and capitalism, as the twin forces of a 'hopeless frenzy of unchained technology and of the rootless organization of the average man' (Heidegger 2000b: 40). For the ideologues of National Socialism, it lay above all in a racial-political form, in 'Judeo-Bolshevism', but also in the 'Asiatic'. In a speech of 19 September 1941, shortly after the launch of Operation Barbarossa, for instance, Hitler described the Ukrainians and the Russians as 'nihilist-asiatic' (quoted in Faye and Cohen-Halimi 2008: 263).

Even at the end of the Second World War, after the defeat of Nazism, German thinkers (including Heidegger) were still prepared to see Western culture in its entirety as essentially nihilist. The deployment of the concept cut across political differences. Whereas Heidegger's critique came from the Right, in his *Farewell to European History, or The Conquest of Nihilism* (1945) Alfred Weber, who had been dismissed from his post at the University of Heidelberg for his criticism of National Socialism, argued that 'we' must

> find our bearings anew and overcome the prevailing Nihilism of our time – that same Nihilism which is the deep-seated cause, as greater minds have already realized, of the historical catastrophe which we of the West, and particularly we Europeans, have brought upon the world. It is as imperative for us to overcome this Nihilism as it is to overcome our old historico-sociological conceptions regarding the possible outward patterns of human life, which were rooted in the now obsolete spatial conditions. We have acted on these conceptions right up to the present as though we were still living on the old earth – with the result that, with this last war, we have laid our life on this world as we have known it hitherto, finally in ruins. (Weber 1947: xi)

From this it is clear that, for Weber, the nihilism responsible for the 'historical catastrophe' is a general phenomenon extending beyond any one philosophy, ideology, or politics.

In contrast, what distinguishes much post-war thinking of nihilism is a tendency to see National Socialism, which had presented itself as essentially anti-nihilist in nature, as not merely one form of nihilism among others but the very consummation of nihilism. The first major work in which National Socialism was identified as a form of nihilism was Hermann Rauschning's *The Revolution of Nihilism* (1938). Nihilism as Rauschning defines it (which is to say, politically rather than philosophically) is 'the total rejection of any sort of doctrine'. National Socialism is, he argues, like Stalinism, a revolution without doctrine, a revolution that 'is nothing but destruction, the dissolution and annihilation of the old elements of public order. It is destroying everything it lays its hands on' (Rauschning 1939: 60, 59). What distinguishes the National Socialist revolution from all others before or contemporaneous with it, is that it cannot tolerate any form of political alterity: 'It is the essential task of every revolution to produce a tabula rasa, to make a clean sweep of the past political forces; but the nihilist revolution of National Socialism sets out to destroy everything that it cannot itself take over and convert to its own pattern' (93). Only as a purely destructive force does it achieve anything: 'National Socialism succeeds everywhere as an element of dissolution, or of disturbance of the existing order, or where it finds fresh material to consume. It fails wherever it attempts any genuine constructive work' (96).

In the post-Second World War era, the view that National Socialism was an – if not the most – extreme form of nihilism became widespread. Leo Strauss, for instance, identified Nazism as 'the most famous form of nihilism' and refined Rauschning's claim regarding the similarity between Nazism and Stalinism by distinguishing between what he took to be the 'brutal' nihilism of Nazism and the 'gentle' nihilism of various Marxist regimes. It is, however, in the work of Theodor Adorno that the conception of National Socialism as a form of nihilism, and of the Holocaust of the consumption of that nihilism, finds what is arguably its most influential expression. Although it contains reflections upon the Nazi 'disfranchising and destruction of German Jewry', which he presents as part of the destruction of Christianity (97), Rauschning's *Revolution of Nihilism* was published before the so-called 'Final Solution' (*Endlösung*) had been officially approved at the Wannsee Conference on 20 January 1942. Adorno, on the other hand, not only thinks nihilism in explicit relation to the Holocaust, but sees what he terms 'Auschwitz' as the consummation of nihilism. As he makes clear in the 1965 lecture series *Metaphysics: Concept and Problems,* by 'Auschwitz' Adorno means not only that particular extermination camp in Poland or even the 'Final Solution' as a whole, but 'the world of torture which has continued to exist after Auschwitz and of which we are receiving the most horrifying reports from Vietnam' (Adorno 2000: 101). A similar view is taken by the French philosopher Philippe Lacoue-Labarthe, who argues in *The Fiction of the Political* (1987) that 'if it is true that the age is that of the accomplishment of nihilism, then it is at Auschwitz that that accomplishment took place in the purest formless form' (Lacoue-Labarthe 1990: 37).

For reasons that no doubt originated as much in his politics as in his philosophy – assuming, *concesso non dato*, that the two are separable – Heidegger avoided almost all public reference to the Holocaust, his one comment on the Nazi extermination camps occurring in a lecture delivered in Bremen on 1 December 1949. In this lecture, Heidegger describes modern agriculture as 'a motorized food industry' that is 'the same thing in its essence as the production of corpses in the gas chambers and the extermination camps' (Heidegger quoted in Lacoue-Labarthe 1990: 34). Crucially, for Heidegger, the full disclosure of what took place in the Nazi extermination camps does not entail a fundamental reorientation of the thinking of Being, this extermination apparently being for him merely one instance among others of man's domination by modern technology as 'Enframing' (*Ge-Stell*). For Adorno, who took Heidegger to be his chief philosophico-political opponent and who sought to challenge his influence most directly in *Jargon of Authenticity* (1964), the fact that Auschwitz took place changes everything, not only for philosophy, but for all of culture, not least the arts, and indeed for human life as such. However, in an irony that runs throughout the history of the deployment of the concept of nihilism, and that is integral to its being what Nietzsche terms 'the uncanniest of all guests', Adorno nonetheless repeats Heidegger in certain key respects in the very act of opposing him.

Like Heidegger, Adorno turns the charge of nihilism back against the very philosopher from whom he inherits it. Just as Heidegger sees Nietzsche's thinking of Being in terms of value and of will to power as the consummation of nihilism, so Adorno sees Heidegger's thinking of Being as nihilism on account of its being profoundly 'hostile to man', centred as it is around 'being towards death and the negating nothingness' (Adorno 1982: 189). Adorno's redetermination of nihilism as that which is 'hostile to man' extends far beyond Heidegger's philosophy, however, to encompass the entire 'administered world' (*verwaltete Welt*) in which human beings now find themselves. It is this administered world of modernity, characterized by reification and the domination of nature, and governed by the principle of identity, that, according to Adorno, reaches its consummation at 'Auschwitz' as what in *Negative Dialectics* (1966) he terms 'absolute integration' (Adorno 1973: 362). This integration is nihilist in that it reduces alterity, difference, and thus life to nothing.

For Adorno, then, the essence of nihilism does not lie in a devaluation of the highest values, or in the forgetting of Being, or in a deracination effected by technology. Rather, it lies in the negation of otherness, the dark triumph of the principle of identity. The only way in which to counter this nihilist destruction of alterity is to remain within the negative, to resist the temptation of the positive. It is precisely this abiding within the negative, this ethically justified resistance to all positivity, that distinguishes Adorno's 'negative dialectics' from the Hegelian dialectic, in which the 'tremendous power' of the negative is orientated towards the positive:

> Hegel's philosophy contains a moment by which that philosophy, despite having made the principle of determinate negation its vital nerve, passes over into affirmation and therefore into ideology: the belief that negation, by being pushed far enough and by reflecting itself, is one with positivity. That, Ladies and Gentlemen, the doctrine of the positive negation, is precisely and strictly the point at which I refuse to follow Hegel. (Adorno 2000: 144)

If one is to resist the nihilism of the positive that is accomplished through the Hegelian negation of the negation, then one has to 'immerse oneself in the darkness as deeply as one possibly can' (144). Such an immersion can be achieved, according to Adorno, only through thought's becoming 'a thinking against itself' (Adorno 1973: 365).

This abiding within the negative might itself seem to be a form of nihilism. Indeed, in the section entitled 'Nihilism' in his major late work, *Negative Dialectics*, Adorno observes that the term has been mobilized by cultural commentators in the post-war era in the interests of the 'moral defamation' of those who refuse to accept 'the Western legacy of positivity and to subscribe to any meaning of things as they exist' (380). Adorno's response to this charge of nihilism directed at his own thought is to reject the possibility of any overcoming (*Überwindung*) of nihilism. His reasoning is that all 'Acts of overcoming – even of nihilism, along with the Nietzschean type that was meant differently and yet supplied fascism with slogans – are always worse than what they overcome' (380). Anticipating the postmodern rethinking of nihilism by the Italian philosopher Gianni Vattimo, to which we shall turn in Chapter 5, Adorno even gestures towards a revalorization of nihilism, claiming that in the world as it is now it is morally justifiable to ask 'whether it would be better for nothing at all to be than something' (380). Rather than trying to defend himself against the charge of nihilism, 'A thinking man's true answer to the question whether he is a nihilist would probably be "Not enough" – out of callousness, perhaps, because of insufficient sympathy for anything that suffers' (380).

This revalorization of nihilism is, however, more apparent than real. Like Heidegger, Adorno continues to use the term 'nihilism' as the ultimate defamation of modernity. The section on nihilism in *Negative Dialectics*, for instance, ends with a distinction between those who, like Adorno himself, have been accused of nihilism for their refusal to assign a positive meaning or purpose to existence and those who make this accusation, it being the latter who are in fact guilty of nihilism: 'The true nihilists are the ones who oppose nihilism with their more and more faded positivities, the ones who are thus conspiring with the extant malice, and eventually with the destructive principle itself. Thought honors itself by defending what is damned as nihilism' (381). In addition to his own thought, Adorno sees that which is 'damned as nihilism' as above all the art of modernist writers such as Franz Kafka and Samuel Beckett. It is no coincidence that Beckett's name should appear in the section on nihilism in *Negative Dialectics*, for Adorno finds in Beckett's work precisely that abiding within the negative that he takes to be the only genuine countering of nihilism.

In his deployment of the concept of nihilism, then, Adorno belongs to a long tradition that repeatedly traverses the political division between Right and Left. Within that tradition, the concept of nihilism is subjected to a series of redeterminations, in each case so that it can be made to operate within a philosophico-political critique of modernity. In each case, this redetermination involves a turning back of the charge of nihilism against that thinker or those thinkers from whom it is appropriated. And in the major cases – Nietzsche, Jünger, Heidegger, Adorno – the counterforce to nihilism comes to be located in a particular form of the aesthetic. In Nietzsche, this counterforce is a Dionysian art that affirms life as a becoming (*Werden*) without unity, purpose, or truth. In Heidegger, it is the *Dichtung* of Friedrich Hölderlin, which names Being as the 'holy' and thus resacralizes it. In Adorno, it is above all the modernism of Kafka and Beckett, to which we shall turn in Chapter 4.

Theorists of nihilism in France: Camus, Blanchot, Cioran

If, in their very different ways, Heidegger and Adorno are the key philosophical figures in the thinking of nihilism after Nietzsche in Germany, Albert Camus is arguably the key figure in France. Like Adorno, Camus redetermines and redeploys the concept from the political Left. In his preface to the 1955 edition of *The Myth of Sisyphus* (originally published in 1942), Camus states that the work was written in 1940 'amidst the French and European disaster', and that it 'declares that even within the limits of nihilism it is possible to find the means to proceed beyond nihilism'. He then goes on to observe that, 'In all the books I have written since, I have attempted to pursue this direction' (Camus 1975: 7). This faith in the possibility of overcoming nihilism remains firm throughout Camus's reflections on the subject in his later works.

In *The Myth of Sisyphus*, Camus offers a vision of the universe as irrational and meaningless, the only certainties being that we cannot know the world and that we will be reduced to nothingness. In this sense, Camus's vision of nihilism is not historical, and is not tied to the experience of modernity. Camus goes on to clarify that the absurdity of the human predicament lies in the 'divorce between the mind that desires and the world that disappoints, my nostalgia for unity, this fragmented universe and the contradiction that binds them together' (50). Experiencing meaninglessness but striving for meaning, experiencing irrationality but striving for reason, the human being is committed – absurdly – to the impossible. Reason is 'ridiculous' in its placing of the human being (as the so-called rational animal) 'in opposition to all creation', given 'the impossibility of reducing this world to a rational and reasonable principle' (51). For Camus, however, this does not mean that all values have been lost. Far from it. For he argues that there is 'integrity' and, indeed, heroism in remaining within the absurd rather than in leaping beyond it in the manner of Kierkegaard (50). The new values that emerge out of this experience of the absurdity of human existence are grounded in revolt; that is, 'the certainty of a crushing fate, without the resignation that ought to accompany it' (54). It is this revolt against the impossible that 'gives life its value' (54). Revolt, then, is that which saves the human being from nihilism by bestowing existence with value.

If Camus's conception of nihilism is essentially Nietzschean, repeating the latter's claim in *The Birth of Tragedy* (1872) that Dionysian wisdom lies in a sense of the horror and absurdity of existence, so too is the privilege he accords to the artist as the one who experiences nihilism at its most extreme and whose response to it is value-producing. According to Camus, the creative artist is the 'most absurd' of human beings, since 'All existence for a man turned away from the eternal is but a vast mime under the mask of the absurd. Creation is the great mime' (87). Like Nietzsche, while championing art as the true counterforce to nihilism (an idea to which a more detailed consideration will be given in Chapters 3 and 4), Camus identifies a particular kind of art as the true counterforce to nihilism, namely the great 'philosophical novelists', among whom he includes Honoré de Balzac, the Marquis de Sade, Herman Melville, Stendhal, Fyodor Dostoevsky, Marcel Proust, André Malraux, and Franz Kafka. In their works, he claims, one finds the 'staggering evidence of man's sole dignity: the dogged revolt against his condition, perseverance in an effort considered sterile' (104). As the title of Camus's work indicates, it is in the mythical figure of Sisyphus, endlessly repeating the act of pushing a gigantic rock to the top of a hill, only for it to roll down the other side, that he finds the 'hero' of the absurd (180), the one who commits himself unreservedly to the endless repetition of the pointless in full awareness of what he is doing. This is Camus's version of Nietzsche's affirmation of eternal recurrence, an affirmation of nihilism that is taken to be the overcoming of nihilism.

If Camus introduces the concept of nihilism in *The Myth of Sisyphus*, it is in his next major philosophico-political work, *The Rebel* (1951), that he charts a history of nihilism in modern European thought that extends back before Nietzsche to the Marquis de Sade, with whom, he argues, 'really begin the history and the tragedy of our times' (Camus 1971: 43). Here, then, Camus explicitly ties the experience of nihilism to that of modernity, and, more precisely, to the other side of the dialectic of modernity as Horkheimer and Adorno will theorize it in *Dialectic of Enlightenment* (1947). If the history of modernity as the experience of nihilism commences with de Sade, it is with Nietzsche that nihilism 'becomes conscious for the first time', for it is Nietzsche who 'recognized nihilism for what it was and examined it like a clinical fact' (57). Like Nietzsche, Camus identifies Dostoevsky as the key figure in nihilism's becoming conscious of itself, nihilism here being defined by Camus as the experience of the loss of absolute values. As Camus sees it, the history of 'contemporary nihilism' begins with the principle that, if there is no God, then everything is permitted, the idea articulated by Ivan Karamazov in Dostoevsky's *The Brothers Karamazov* (1879–80) (see Camus 1971: 52). The nihilist response to this 'Everything is permitted' is not simply despair, but 'the *desire* to despair and to negate' (52; emphasis added).

Just as Nietzsche distinguishes between two basic forms of nihilism – active and passive – so Camus distinguishes between two forms of contemporary nihilism: individual and state. The former he finds epitomized by nineteenth-century Russian nihilism and, in the twentieth century, by communism and Nazism. Camus identifies Nechaev as the Russian nihilist whose thought most fully anticipates the nihilism of Nazism, since it is Nechaev who 'pushed nihilism to the farthest point' (129). By this, Camus means that, in Nazism, the will to negate is at its most extreme: Nazism is not merely suicidal, but the incarnation of the will to an 'absolute destruction, of both oneself and everybody else' (15). Like Adorno, then, Camus sees Nazism as the consummation of the nihilism of modernity: 'The crimes of the Hitler régime, among them the massacre of the Jews, are without precedent in history because history gives no other example of a doctrine of such total destruction being able to seize the levers of command of a civilized nation' (153).

In direct opposition to both state and individual nihilism, Camus sets 'the rebel' (*le revolté*). The key difference between the rebel and the nihilist is that the former aims as justice, truth, and happiness, whereas the latter increases injustice and falsehood, and 'destroys, in its fury, its ancient demands and thus deprives rebellion of its most cogent reasons' (249). Of course, if nihilism is taken seriously, then those very values to which the rebel appeals – justice, truth, and happiness – are themselves without universal validity, and it is here that Camus's appeal to the concept of the rebel as the answer to the experience of nihilism shows its weakness, since, unlike Nietzsche, Camus does not attempt any revaluation of all values.

It is upon the weaknesses besetting Camus's attempts to propose an escape from nihilism in *The Rebel* that another key figure in the theorization of nihilism in post-war France focuses. Like Camus, Maurice Blanchot (1907–2003) was both a theorist and a writer of fiction, although, unlike Camus, Blanchot moved from a position on the nationalist Right in the 1930s to the Left in the post-war era. Blanchot's engagement with the concept of nihilism in its relation to the literary commences in the opening section of his first collection of literary-critical writings, *Faux Pas* (1943), and recurs at key moments in works published over the following four decades. His most sustained engagement with the concept of nihilism is to be found in two essays on Camus published in 1954 under the titles 'Reflections on Hell' and 'Reflections on Nihilism' and an essay on Nietzsche first published in 1958 under the title 'Crossing the Line', these three essays being included in Blanchot's 1969 collection, *The Infinite Conversation*. While approving of Camus's vision of the absurd in *The Myth of Sisyphus*, Blanchot takes his distance from Camus's position in *The Rebel* precisely on account of its central argument that in the figure of the rebel the step beyond nihilism is taken. According to Blanchot, what Camus fails to grasp is that 'It may be, in fact, that what we call nihilism has been at work in this obscure constraint that turns us away from it; that it was the very thing that hides it, the *movement of detour* making us believe we have always already put nihilism aside' (Blanchot 1993: 177; Blanchot's emphasis). This view of nihilism stands in striking contrast to Blanchot's own dismissal of what he describes as the 'naïve calculation' of the nihilist in *Faux Pas*, and leads Blanchot to conclude that 'one of the errors of *The Rebel* is that, on the pretext of rapidly putting nihilism out of play, it in reality plays into nihilism's hands by accepting its own self-effacement, the autodisappearance that is but its visage and the seduction this visage exercises' (180).

As Leslie Hill observes, Blanchot's own conception of nihilism as articulated in his 1958 essay on Nietzsche is distinct from Camus's above all in its movement away from the question of values (and their loss) to that of possibility: 'The crux [...] in the debate concerning nihilism for Blanchot is not that supposed loss of higher, transcendent values which Heidegger adopts as his starting point, but rather the expectation, embodied in a certain view of the figure of the *Übermensch*, that all is possible' (Hill 2001: 242). In short, for Blanchot the essence of nihilism lies not in Ivan Karamazov's 'Everything is permitted' but in the idea that 'Everything is possible'. Blanchot thinks the limit of nihilism as the impossibility of that which, according to Heidegger, is the ownmost possibility of the human being, namely death. As Heidegger puts it in *Being and Time* (1927), 'Death is the possibility of the absolute impossibility of Dasein. Thus death reveals itself as that *possibility which is one's ownmost, which is non-relational, and which is not to be outstripped*' (Heidegger 1962: 294; Heidegger's emphasis).

Heidegger's centrality to the thinking of nihilism in the 1950s was owing to essays published in *Off the Beaten Track* (1950) and *Lectures and Essays* (1954), and especially to 'On the Question of Being' (1955), in which (as we have seen above) Heidegger responds to Jünger's *Across the Line* (1950), arguing that there can be no overcoming (*Überwindung*) of nihilism in either Nietzsche's or Jünger's sense of the term. While, as Hill notes, there are certainly important differences between Blanchot's and Heidegger's thinking of nihilism, Blanchot nonetheless shares Heidegger's post-war position on the overcoming of nihilism. Indeed, Blanchot even defines nihilism itself as 'the possibility of all going beyond [*dépassement*]' (Blanchot 1993: 145). As we shall see in Chapter 5, when we consider the relation between postmodernism and nihilism, this sense that there can be no overcoming of nihilism – a view that, for all their differences, is shared by the Heidegger of the post-war era, by Adorno, and by Blanchot – will prove to be crucial in the revalorization of nihilism in certain strains of postmodern thought.

In the figure of Sisyphus as presented by Camus, Blanchot finds an example of the experience of radical impossibility; that is, the impossibility of death: 'Sisyphus is the approach to this region where even the one who commits suicide by an act that is personal and a will that is resolute collides head-on with death as with a density no act can penetrate and that cannot be proposed as an end or a goal' (179). According to Blanchot, this is the experience of a region that is 'approached in life by all who, having lost the world, move restlessly *between* being and nothingness: a swarming mass of inexistence, a proliferation without reality, nihilism's vermin: ourselves' (179). It is just such an experience of inexistence that Blanchot will place at the heart of his own fictional writings, in works such as *Death Sentence* (1948), *The Madness of the Day* (1949), and *The Instant of My Death* (1994), whose relation to nihilism will be considered in Chapter 4. That this experience is presented by Blanchot as *both* the experience of nihilism *and* that which resists nihilism indicates how the uncanniness of nihilism to which Nietzsche refers can manifest itself in the work of those who seek to master the concept in the interests of critique.

In Camus and Blanchot, then, one finds an engagement with the concept of nihilism that in two key respects belongs within a tradition that commences with Nietzsche. Both writers characterize modernity in terms of a nihilism that is to be seen as the greatest of threats, and, as we shall see in Blanchot's fictional work when we turn to it in Chapter 4, both writers privilege the literary as that which constitutes the true counterforce to nihilism. Aside from Camus, the other major thinker in France who is generally taken to have engaged most directly with the question of nihilism in the post-Second World War period was the Romanian-born philosopher E. M. Cioran (1911–95). After having published five books in Romania in the 1930s – including *On the Heights of Despair* (1934), *The Book of Delusions* (1936), and *The Transfiguration of Romania* (1937) – Cioran moved to France in 1937, and published his first work in French, *A Short History of Decay*, in 1949. This was followed by a series of other works in French over the next four decades, including *The Temptation to Exist* (1956), *History and Utopia* (1960), and *The Trouble with Being Born* (1973). Cioran's thought is deeply indebted to Nietzsche's, above all in his critique of Enlightenment values and in his characterization of modernity as decadent. Cioran is no less indebted to the style of Nietzsche's thought than he is to its content, much of his writing taking the form of essays and aphorisms, and avoiding technical philosophical terminology. However, while Cioran may follow Nietzsche in rejecting the key values of the Enlightenment (above all, the power of reason and the concept of progress), and if the emphasis falls, as the titles of his works indicate, upon despair, decay, mortality, finitude, and the nothing, Susan Sontag is no doubt right to insist that what distinguishes Cioran's thought from Nietzsche's is anything akin to the latter's 'heroic effort to surmount nihilism' (Sontag in Cioran 1987: 26). In *The Trouble with Being Born*, for instance, Cioran declares that he finds Nietzsche 'too *naïve*', since he 'demolished so many idols only to replace them with others' (Cioran 1993: 85).

The conclusion that Cioran's position is, from the outset, a nihilist one might seem perfectly justified by the following passage, from a section entitled 'Nothing Matters' in his first book, *On the Heights of Despair*, in which he connects the loss of values that will be Camus's concern with the 'Everything is possible' that will be at the heart of Blanchot's conception of nihilism:

> Everything is possible, and yet nothing is. All is permitted, and yet again, nothing. No matter which way we go, it is no better than any other. It is all the same whether you achieve something or not, have faith or not,

just as it is all the same whether you cry or remain silent. There is an explanation for everything, and yet there is none. Everything is both real and unreal, normal and absurd, splendid and insipid. There is nothing worth more than anything else, nor any idea better than any other. (Cioran 1992: 116)

In his post-war works, Cioran repeatedly characterizes the present age as decadent, caught in a process of irreversible decline that can be compared to those which befell ancient Greece and Rome. Modernity has reached a point where its values are exhausted, and yet no new values have been generated. Perhaps surprisingly, however, the term 'nihilism' occurs very rarely in Cioran's work. When it does, it is presented as that which is to be avoided. In the chapter entitled 'Faces of Decadence', in *A Short History of Decay*, for instance, Cioran argues in Nietzschean fashion that 'Decadence is merely instinct gone impure under the action of consciousness. Hence we cannot overestimate the importance of gastronomy in the existence of a collectivity. The *conscious* act of eating is an Alexandrian phenomenon; barbarism *feeds*' (Cioran 1990: 112; Cioran's emphasis). He then goes on to claim that:

> To meditate upon one's sensations – to *know* one is eating – is an accession of consciousness by which an elementary action transcends its immediate goal. Alongside intellectual disgust develops another, deeper and more dangerous: emanating from the viscera, it ends at the severest form of nihilism, the nihilism of repletion. (112–13)

And in *Syllogisms of Bitterness* (1952), he observes: 'A little more fervor in my nihilism and I might – gainsaying *everything* – shake off my doubts and triumph over them. But I have only the taste of negation, not its grace' (Cioran 1999: 31; Cioran's emphasis).

Overall, rather than identifying himself as the nihilist that some commentators have suggested he is, Cioran seeks to face up to the threat of nihilism, but not in order to overcome it. Like other thinkers on the political Right such as Ernst Jünger and Martin Heidegger – and in the 1930s and early 1940s Cioran supported the fascist Iron Guard in Romania – he came in the post-war period to reject the possibility of overcoming nihilism. However, this rejection does not, as he sees it, make of Cioran himself a nihilist. Indeed, he insists that he is 'not a nihilist', even though he was 'always tempted by negation' (Cioran 1995: 221–2; my translation). This position is supported by statements on negation such as the following, in *The Temptation to Exist*:

> It is true that negation is the mind's first freedom, yet a negative habit is fruitful only so long as we exert ourselves to overcome it, adapt it to our needs; once *acquired* it can imprison us – a chain like any other. And slavery for slavery, the servitude of existence is the preferable choice, even at the price of a certain self-splintering: it is a matter of avoiding the contagion of nothingness, the comforts of the abyss … (Cioran 1987: 207; Cioran's emphasis)

Cioran's position is ultimately closer to scepticism than it is to nihilism. By scepticism here is to be understood an undoing of values that never arrives at their complete annihilation. As Cioran puts it in the essay 'Skeptic and Barbarian', in *The Fall into Time* (1964): 'Every affirmation and, with all the more reason, every belief proceeds from a depth of barbarism which the majority, which virtually the totality of men have the good fortune to preserve, and which the Skeptic [...] has lost or liquidated, so that he retains only vague vestiges of it, too weak to influence his behavior or the conduct of his ideas' (Cioran 1970: 89). As becomes clear in his remarks on Dostoevsky in *History and Utopia*, nihilism as Cioran understands it is the step beyond scepticism, the annihilation of any vestige of belief or value (see Cioran 1996: 112), and, for all his attempted liquidation of Enlightenment values, in his post-war work Cioran does retain the vestiges of those values.

Cioran's relation to nihilism is akin to the one he envisages for Samuel Beckett, to a consideration of whose work we shall turn in Chapter 4 and of whom Cioran wrote in 1976:

> Ever since our first encounter, I have realized that he reached *the limit*, that he perhaps began there, at the impossible, at the exceptional, at the impasse. And the admirable thing is that he has not *budged*, that having come up against a wall from the start, he has persevered, as valiant as he has always been: the limit-situation as point of departure, the end as advent! (Cioran 1991: 134–5; Cioran's emphasis)

These remarks on Beckett gesture towards a resistance of nihilism that never becomes a naïve dream of overcoming. This attitude to nihilism places Cioran in a tradition that includes a number of key thinkers who, in the 1930s, not only characterized modernity as nihilist but also believed in the possibility of a right-wing political overcoming of that nihilism, and who in the post-war era came to take up positions in which any overcoming of nihilism was rejected, and the resistance of nihilism was rethought in another form, namely as endurance. Nietzschean in inspiration, although taking its distance from the programmatic side to his philosophical modernism, this tradition includes Jünger, Heidegger, Blanchot, and Cioran. That each of them should think the literary as a way of resisting nihilism without falling for the lure of overcoming is far from insignificant. And, as we shall see in the next two chapters, this turn to the literary is in fact no less Nietzschean than are anxieties about the nihilism of modernity.

Part II

AESTHETIC MODERNISM AND NIHILISM

3

FROM FLAUBERT TO DADA

In the previous two chapters, we have seen that, commencing with Nietzsche, each of the major philosophical deployments of the concept of nihilism against a variously defined modernity includes a theorization of a certain form of the aesthetic as that which either resists or overcomes nihilism. As we shall see in the next two chapters, however, not only is aesthetic modernism a complex phenomenon in itself, but the very ambiguity that, according to Nietzsche, characterizes nihilism as the 'uncanniest of all guests' is reflected in debates concerning the relationship between nihilism and aesthetic modernism. By aesthetic modernism here is to be understood a broadly conceived artistic movement that, within the sphere of the literary, extends from mid-nineteenth-century French writers such as Charles Baudelaire and Gustave Flaubert, through the European avant-garde movements of the early decades of the twentieth century, in particular Dada, to the 'high' modernism of Franz Kafka and the 'late' modernism of Samuel Beckett and Paul Celan. On the one hand, aesthetic modernism is repeatedly seen, and indeed repeatedly sees itself, as the counterforce to the perceived nihilism of modernity. On the other hand, it is also seen, and on occasion sees itself, as the incarnation of nihilism. Furthermore, like the philosophical deployment of the concept of nihilism, the interpretation of aesthetic modernism as either the counterforce to, or the incarnation of, nihilism cuts across political borders in ways suggesting that, where the question of nihilism arises, it is always a matter of borders, limits, and extremes, both the threat and the promise of that which inhabits the liminal space between the known and the unknown.

The 'tortured romantic' and the 'well-made phrase': Bourget on Flaubert

The most influential charge of nihilism directed explicitly against a particular form of aesthetic modernism occurs in the French critic and novelist Paul Bourget's two-volume *Essays in Contemporary Psychology* (1883 and 1885). As Matei Calinescu observes, Bourget is the first French writer 'to accept unwaveringly (unlike Baudelaire or even Gautier) both the term and the fact of decadence, and to articulate this acceptance in a full-blown, philosophic and aesthetic theory of decadence as a style' (Calinescu 1987: 169). Through an analysis of key works by ten major writers (Baudelaire, Leconte de Lisle, Flaubert, the Goncourts, Dumas *fils*, Renan, Taine, Stendhal, Turgenev, and Amiel), Bourget aims to show how a 'sickness' entered the 'moral life' of France during the second half of the nineteenth century. This sickness takes the form of a profound and continuous 'pessimism', a 'spirit of negation and depression' (Bourget 1912: i. xxi–xxii; my translation). Among the causes of this spiritual sickness, Bourget includes dilettantism (Renan and the Goncourt brothers), cosmopolitan life (Stendhal, Turgenev, and Amiel), the perversions of love (Baudelaire and Dumas), the effects of science (Flaubert, Leconte de Lisle, and Taine), and the conflict between democracy and high culture (Renan, the Goncourts, Taine, and Flaubert). To these, Bourget adds two key historical events, both of them 'social tragedies': the Franco-Prussian War of 1870–1 and the Paris Commune of 1871.

Although, in the preface to the 1885 edition of his *Essays*, he refers to the 'nihilism' of Maupassant – the word 'nihilism' having been appropriated from Prosper Mérimée's 1865 French translation of Turgenev's novel *Fathers and Sons* (1862) – Bourget uses the term principally of the works of Gustave Flaubert. According to Bourget, Flaubert's particular brand of nihilism is that of a 'tortured romantic' (148), his works reflecting the collision of a romantic and a scientific outlook, and presenting the human being as alienated from a reality that remains nonetheless 'ineluctable' (155). According to Bourget, the novels *Madame Bovary* (1857) and *The Sentimental Education* (1869) offer unremittingly pessimistic visions of the alienating effects of literature on the mind, whereas the unfinished and posthumously published *Bouvard and Pécuchet* (1881) presents the devastating effects of science. Bourget's characterization of Flaubert as nihilist was anticipated by Emile Zola, who, in his chapter on Flaubert in *The*

Naturalist Novelists (1881), argues that, alongside the 'impeccable stylist', there is a 'philosopher' who is 'the most extensive negator that we have ever had in our literature. He professes true nihilism [*le véritable nihilisme*]' (Zola 1923: 196; my translation).

Bourget's interpretation of Flaubert and other mid-nineteenth-century French writers is repeated in the twentieth century by the German-Jewish philosopher Karl Löwith. In *European Nihilism: Reflections on the Spiritual and Historical Background of the European War* (1940), Löwith argues that 'Around the middle of the century, nihilism found its most significant expression in Flaubert and Baudelaire' (Löwith 1995: 193). Löwith sees this tradition continuing into the twentieth century, with a series of major writers offering profoundly nihilist visions of modern humanity:

> To envision the Nothing of modern humanity by employing all the means of art and spirit, had also been the task of the writer, in whose works the possibilities of the novel are exhausted. None of them give shape any more to an authentically human world; they simply analyze intellectual developments, mental reactions, and social relationships. Marcel Proust and André Gide, Thomas Mann and Aldous Huxley, André Malraux and D. H. Lawrence, Joyce and Céline – none of them give shape any more, as do the great novels from Cervantes to Dickens and from Balzac to Tolstoy, to a human cosmos; they simply convey a disheartening truth about human beings, in connection with which the human being as such disappears. (197)

While Löwith follows Bourget in his identification of the origins of nihilism in literature, there is one crucial difference between their respective positions. Unlike Löwith, Bourget takes only the *content* of Flaubert's works to be nihilist, arguing that in *literary style* – the 'well-made phrase' (Bourget 1912: ii. 170) – Flaubert finds a form of resistance to the nihilism resulting from the collision of romantic and scientific outlooks. In this respect, Bourget's Flaubert anticipates some of the major writers of the twentieth century as seen through a critical discourse that not only deploys the concept of nihilism but also repeatedly identifies the aesthetic as either an expression of, or a privileged form of resistance to, nihilism. This is the side to Flaubert's project on which Roger Griffin places the emphasis when he argues that the French writer's oeuvre is 'the fulfilment of a self-appointed mission to weave a new sacred canopy out of the poetic, world-creating, power of language and so transcend the "dead time" that was corrupting society from within' (Griffin 2007: 91). Any such conception of the aesthetic as saving has, however, to be placed in relation to repeatedly expressed anxieties which, on occasion, extend as far as the outright condemnation of aesthetic modernism as the consummation of nihilism.

The 'only superior counterforce': Nietzsche on art versus nihilism

Bourget's critique of an entire French literary tradition as nihilist is significant less for any intrinsic value it might possess than for its influence on others, and above all on Nietzsche. In *Twilight of the Idols* (written in 1888, published in 1889), for instance, Nietzsche follows both Zola and Bourget in characterizing Flaubert as nihilist, although in a manner that is considerably more playful than theirs: '*On ne peut penser et écrire qu'assis* (G. Flaubert). – I've caught you, nihilist! Sitting still is the very *sin* against the Holy Spirit. Only *peripatetic* thoughts have any value' (Nietzsche 2005: 160; Nietzsche's emphasis). However, while Nietzsche certainly believes in the possibility of a nihilistic art – for which another term will be 'romantic art' – an entry in his May–June 1888 notebook suggests that he takes art as such to be the privileged form of resistance to nihilism: 'Art as the only superior counterforce to all will to denial of life, as that which is anti-Christian, anti-Buddhist, antinihilist *par excellence*' (Nietzsche 1968: 452, 1999b: xiii. 521). In order to understand both why Nietzsche should come to consider art to be such a counterforce to nihilism, and what kind of art he is thinking of here, one has to return to his most extensive engagement with the theory of art, in his first book, *The Birth of Tragedy out of the Spirit of Music* (1872).

In his original preface to *The Birth of Tragedy*, addressed to the composer Richard Wagner, Nietzsche expresses on more than one occasion his conviction that 'art represents the highest task and the truly metaphysical activity of this life' (Nietzsche 1967: 31–2). But why should art be the highest task? What makes it the metaphysical activity *par excellence*? In his 1870 essay 'The Dionysiac World View', which remained unpublished during his lifetime but which anticipates in certain key respects the central argument of *The Birth of Tragedy*, Nietzsche outlines his conception of the nature and function of Greek art, arguing that it served 'to perfect existence, to augment it and seduce men into continuing to live' (Nietzsche 1999a: 125). The Greeks, he claims, 'knew the terrors and horrors of existence, but they covered them with a veil in order to be able to live' (124). As in *The Birth of Tragedy*, so in this essay, Nietzsche's shifts between conceiving of art on the one hand as a veil that hides the horror of existence, and on the other as that which transfigures reality: 'To view its own existence in a transfiguring mirror [*in einem verklärenden Spiegel*] and to protect itself with this mirror against the Medusa – this was the genial strategy adopted by the Hellenic "Will" in order to be able to live at all' (125).

In *The Birth of Tragedy* itself, Nietzsche argues that the true nature of existence is glimpsed only by 'Dionysian man', who 'resembles Hamlet: both have looked truly into the essence of things, they have *gained knowledge*, and nausea inhibits action [...]. Conscious of the truth he has once seen, man now sees everywhere only the horror or absurdity of existence. [...] he understands the wisdom of the sylvan god, Silenus: he is nauseated' (Nietzsche 1967: 60; Nietzsche's emphasis). The wisdom to which Nietzsche is referring here is to be found in Sophocles' *Oedipus at Colonus*: '"What is best of all is utterly beyond your reach: not to be born, not to *be*, to be *nothing*. But the second best for you is – to die soon"' (42; Nietzsche's emphasis). In a vision derived from Arthur Schopenhauer's *The World as Will and Representation* (1818), Nietzsche sees this horror and absurdity as lying in the fact that the individual exists merely at the level of the phenomenon, while at the level of true being there is no individuation, but rather the primal Will. Art of the highest kind is grounded in the Dionysian experience of this 'reality of nature' (*Naturwirklichkeit*), in which the individual not only lacks any value but has no being at all.

According to Nietzsche, in Greek art at its peak – that is, in Attic tragedy, and, above all, in the works of Aeschylus and, to a slightly lesser extent, Sophocles – this truth about existence is rendered bearable. The 'highest and, indeed, the truly serious task of art' is, he argues, 'to save the eye from gazing into the horrors of night and to deliver the subject by the healing balm of illusion from the spasms of the agitations of the will' (118). Salvation is achieved through a countering of the Dionysian experience by the Apollonian, this latter being the second of the two 'primordial artistic drives'. If the Dionysian drive involves an experience of intoxication (*Rausch*) in which the individual is dissolved back into the primal unity, and finds its purest expression in the art of music, the Apollonian drive involves an experience of dream (*Traum*), of fixed and coherent being, and finds its purest expression in the art of sculpture and also in the image more generally. Although they stand 'in tremendous opposition' to each other, in Attic tragedy these two primordial artistic drives are brought together: 'by a metaphysical miracle of the Hellenic "will", they appear coupled with each other, and through this coupling ultimately generate an equally Dionysian and Apollinian [*sic*] form of art' (33). It is owing to this coupling of the Dionysian and the Apollonian that Attic tragedy can grant us 'metaphysical comfort', making of art a 'saving sorceress, expert at healing' (59–60).

Nietzsche asserts that, through its Dionysian element, Attic tragedy captures the truth of the horror and absurdity of existence, while the Apollonian element submits this experience to a veiling (*Umschleierung*) that renders this truth bearable by making it beautiful. Nietzsche also refers to this veiling as a transfiguration (*Verklärung*), and, while he continues in his later work to describe art as both veiling and transfiguring, it is the latter idea that will come to dominate his thinking of art. It is as a transfiguration of reality that art constitutes an overcoming (*Überwindung*) of the horror and absurdity of existence. Art is 'not merely imitation of the reality of nature but rather a metaphysical supplement of the reality of nature, placed beside it for its overcoming. The tragic myth, too, insofar as it belongs to art at all, participates fully in this metaphysical intention of art to transfigure [*dieser metaphysischen Verklärungsabsicht*]' (140).

Nietzsche sees the chorus as the element in Attic tragedy that serves to articulate the Dionysian experience of the dissolution of the individual subject, and the dramatic dithyramb, combining music and words, as the form of artistic expression that most fully brings the Dionysian and the Apollonian into 'mysterious union' (47). For Nietzsche, the dithyramb will remain the highest form of artistic expression, and he himself writes dithyrambs when he seeks to overcome the distinction between philosopher and poet in his writings of the 1880s, above all in *Thus Spoke Zarathustra* (1883–5). By casting a 'veil of beauty' or a 'splendid illusion' over the 'dissonance' of the Dionysian, the Apollonian element in art makes life 'worth living' (143). On the one hand, this suggests that one of the key definitions of what Nietzsche will later describe as 'nihilism' – that is, a sense of the meaninglessness of existence – lies in the Dionysian experience, and that this is countered by the Apollonian. On the other hand, Nietzsche comes to see the most extreme form of nihilism as

Christianity, which he characterizes as a veiling of reality; this suggests that the Apollonian element is itself nihilist. This ambiguity remains at the heart of Nietzsche's later thinking of art in its relation to nihilism: in short, art becomes for Nietzsche *both* the 'superior counterforce' to nihilism *and* the most extreme form of nihilism.

In the second half of *The Birth of Tragedy*, Nietzsche argues that the 'death of tragedy' – which is to say, the death of genuine art – occurs when the Dionysian–Apollonian opposition is supplanted by the opposition between the Dionysian and the Socratic. According to Nietzsche, Socrates is a new type of humanity: 'theoretical' as opposed to creative-artistic. The 'theoretical' man believes in reason, in the possibility of explaining nature, in knowledge or science (*Wissenschaft*) rather than art as that which renders life bearable. In the tragedies of Euripides, this Socratic view begins to influence artistic production. Crucially, Nietzsche sees the coming to dominance of 'theoretical' humanity as 'the one turning point and vortex of so-called world history' (96). In other words, world history falls into two basic phases: the properly Greek age (the age of tragic art) and the modern age (the age of science). Although he places them in opposition to each other, Nietzsche nonetheless conceives of both art and science in terms of their power to bestow meaning and value on life. Thus they are both thought in terms of what in his work of the later 1880s he will identify as 'nihilism'. Already in *The Birth of Tragedy*, Nietzsche presents art as the only genuine counterforce to an experience that (although he has yet to use the term) bears some of the key characteristics of nihilism. Science, on the other hand, is not merely a failed counter-movement to nihilism, but itself nihilism.

The view that art is the only genuine counterforce to nihilism finds explicit expression in the late work *On the Genealogy of Morals* (1887). There are, however, at least two key differences between Nietzsche's early and later position regarding art. First, he attempts to move away from a Schopenhauerian conception of art as a form of consolation (*Trost*) – this move Nietzsche seeks to achieve by privileging the Dionysian over the Apollonian. Secondly, he changes his mind on where that Dionysian art is realized in modernity. In *The Birth of Tragedy*, Nietzsche's call for a rebirth of tragedy – and of the mythic thinking that lies at its heart – in the modern age is accompanied by an identification of the one artist in whom he sees such a rebirth being accomplished, namely the German composer Richard Wagner. Four years after *The Birth of Tragedy*, in the fourth of his 'Untimely Meditations', entitled *Richard Wagner in Bayreuth* (1876), Nietzsche presents Wagner as the embodiment of the dithyrambic artist in the modern period. The clear binary distinction between Greek and modern culture remains firmly in place in this essay, with the latter culture being subjected to withering critique. Of modern art, Nietzsche asserts that its function is 'stupefaction or delirium! To put to sleep or to intoxicate!' (Nietzsche 1997: 220). What Wagner offers, on the other hand, is a dithyrambic art in which tragic myth finds its true home. The stakes could not be higher, with tragic art alone being identified as that which can save humanity: 'There is only one hope and one guarantee for the future of humanity: it consists in his *retention of the sense for the tragic*' (213; Nietzsche's emphasis). In his 1876 essay, Nietzsche identifies Wagner's *Tristan and Isolde* (1865) as 'the actual *opus metaphysicum* of all art', and *The Ring of the Nibelung* (1869–76) as 'a tremendous system of thought without the conceptual form of thought' (232, 236).

This view of Wagner's art undergoes not merely revision but a complete reversal in Nietzsche's later thought. Rather than being the rebirth of a properly Dionysian art, Wagner's operas come to be seen as the epitome of 'romantic' art, which is to say 'nihilist' art, with Wagner's final opera, *Parsifal* (1882), as the consummation of nihilism in art. Nietzsche's later position on art is laid out in summary form in his preface to the 1886 edition of *The Birth of Tragedy*, under the title 'Attempt at a Self-Criticism'. Here, he recasts the argument of his first book in terms of an opposition not between art and science, but between art and morality, the aesthetic and the moral interpretation of phenomena. Art is now presented as that which stands against Christianity, the latter being characterized as a 'craving for nothingness' (Nietzsche 1967: 23). True (Dionysian) art is a 'fundamentally opposite doctrine and valuation of life' to the Christian/moral one, an affirmation rather than a negation of life. Whereas, in *The Birth of Tragedy* itself, Nietzsche argues that modern German music (Beethoven and, above all, Wagner) is the rebirth of Dionysian art, the closest modern art to that of the Greeks, in the new preface he completely reverses this position, arguing that modern German music is 'romanticism through and through and the most un-Greek of all possible art forms' (25).

In *On the Genealogy of Morals*, Nietzsche labels *Parsifal* 'nihilistic' (Nietzsche 1989: 101). Such art is, he claims, the greatest 'corruption' of art, since it turns the counterforce to nihilism into nihilism (Christian, moral, ascetic, anti-life, anti-sensual, redemptive, and renunciatory). In this late work, genuine (non-Wagnerian) art is presented as that which is diametrically opposed to nihilism as Christianity and science (*Wissenschaft*), Nietzsche here bringing together his idea of art as both life-affirming and deceiving. In art, he declares: 'the *lie* is sanctified and the *will to deception* has a good conscience', and this makes art 'much more fundamentally opposed to the ascetic ideal than is science'. It was, he continues, for precisely this reason that Plato excluded the artist from his Republic:

> Plato versus Homer: that is the complete, the genuine antagonism – there the sincerest advocate of the 'beyond', the great slanderer of life; here the instinctive deifier, the *golden* nature. To place himself in the service of the ascetic ideal is therefore the most distinctive *corruption* of an artist that is at all possible; unhappily, also one of the most common forms of corruption; for nothing is more easily corrupted than an artist. (153–4; Nietzsche's emphasis)

So, while Nietzsche reverses his position on Wagner's art, his early and late work nonetheless share the conviction that true (Dionysian) art is that which stands opposed to the nihilism of modernity in the forms of science (or a 'theoretical' culture, as it is described in *The Birth of Tragedy*) and Christianity (in *On the Genealogy of Morals*). And his conception of the artistic form that is most properly Dionysian also remains in place, for Nietzsche continues to argue that the dithyramb is the most Dionysian of artistic forms. Indeed, in a letter of 11 February 1883 to Franz Overbeck, he states that his new book – *Thus Spoke Zarathustra* – 'is poetry, and not a collection of aphorisms' (Nietzsche 1969: 207), and the work includes what Nietzsche identifies as dithyrambs. In *Ecce Homo* (written in 1888), he singles out 'The Night Song' in the Second Part of *Zarathustra* as a dithyramb expressing tragic wisdom. This dithyramb ends:

> It is night: alas that I must be light! And thirst for the nocturnal! And loneliness!
> It is night: now my longing breaks out of me like a well – I long to speak.
> It is night: now all fountains speak more loudly. And my soul too is a fountain.
> It is night: only now all the songs of the lovers awaken. And my soul too is the song of the lover.
>
> (Nietzsche 2006: 83)

Of this dithyramb, Nietzsche makes the rather dubious claim that 'Nothing like this has ever been composed, ever been felt, ever been *suffered* before: this is how a god suffers, a Dionysus' (Nietzsche 2005: 133; Nietzsche's emphasis). The dithyrambs of Zarathustra were followed by a sequence of nine poems written between 1883 and 1888, which Nietzsche collected together in 1888 under the title *Dithyrambs of Dionysus* (published in 1892).

There is considerable irony in Nietzsche's privileging of art as that which stands opposed to nihilism while dismissing what he considers to be nihilist art as 'romantic', given that his own conception of art is so close to the one articulated in German romanticism, and above all to that of the poet Friedrich Hölderlin (1770–1843). Not only does Hölderlin anticipate Nietzsche's evaluative distinction between Greek and modern culture, but he also accords to poetry the ultimate privilege. The impact of Nietzsche's theory of art coincided in the early decades of the twentieth century with the rediscovery of Hölderlin's work, which was re-edited by Norbert von Hellingrath (1888–1916), the first volume appearing in 1913. It was from this double impact that another major philosophical engagement with art that made it the privileged countermovement to nihilism came about in 1930s Germany.

Literature and *Dichtung*: Heidegger on poetry and the naming of Being

Whereas Nietzsche's first book publication takes art as its subject, and presents a certain form of art (Attic tragedy) and its 'rebirth' in modern German music as the counterforce to nihilism, Martin Heidegger's public engagement with the nature of art came only after he had withdrawn from direct political participation in the Nazi movement. Having resigned as rector of Freiburg University in 1934, Heidegger proceeded in the next two decades to write both on art in general – most notably in *The Origin of the Work of Art* (1935–6) – and on a number of major German poets: Stefan George, Rainer Maria Rilke, Georg Trakl, and, above all, Friedrich Hölderlin. As we have seen in the previous chapter, Heidegger reconceived nihilism against Nietzsche as the forgetting of Being (*Seinsvergessenheit*). In poetry (*Dichtung*), which he distinguishes rigorously from literature (*Literatur*), Heidegger finds a language that not only recollects but also names Being, and thus counters the nihilism of its forgetting. Like Nietzsche before him, then, and beyond all the differences between their conceptions of both art and nihilism, Heidegger comes to see art – in the form of *Dichtung* – as the superior counterforce to nihilism.

In *The Origin of the Work of Art*, Heidegger argues that the genuine work of art 'opens up, in its own way, the being of beings' (Heidegger 2002: 19). Focusing on an unidentified painting by Van Gogh of some peasant shoes, he insists that the artwork is to be understood not as a representation, copy, or imitation of reality, but rather as that through which the being of what is shown is disclosed. In the case of the shoes painted by Van Gogh, the work of art lets the viewer know 'what the shoes, in truth, are' (15). This disclosure of the being of beings is to be understood as unconcealment (*Unverborgenheit*), or what for the Greeks was *aletheia*, 'truth'. Just as Nietzsche claims that the dithyramb is the true form of Dionysiac art, and himself seeks to practise this art in *Zarathustra* and the *Dithyrambs of Dionysus*, so Heidegger claims that *Dichtung* is 'the essence of all art', it being 'the saying of the unconcealment of beings' (45–6), the saying of beings in their truth, and thus the countering of nihilism defined as the oblivion or forgetting of Being.

At its purest, *Dichtung* not only says the being of beings, but is itself about the essence of poetry as such a saying. In the poetry of Hölderlin, Heidegger finds the supreme instance of this kind of art, making of him what in the lecture 'Hölderlin and the Essence of Poetry' (1936) Heidegger terms '*the poet's poet*' (Heidegger 2000a: 52; Heidegger's emphasis). That the original formulation of this thought, prior to its publication, was not 'the poet's poet' but 'the Germans' poet' makes it clear that, just as for the Nietzsche of *The Birth of Tragedy* it is 'the German spirit' that lies at the heart of his thinking of art in its relation to modernity, so for Heidegger genuine poetry is to be thought as *German* poetry, within a distinction between Greek and modern culture closely akin to Nietzsche's, and a similar conviction that it is in modern *German* art that the overcoming of the nihilism of modernity is to be achieved. However, whereas Nietzsche reverses his position on Wagner, Heidegger continues to hold to the conviction that Hölderlin is the most authentic poet of modernity, the only poet comparable to Sophocles in his naming of the being of beings. The reasons for this privileging of Hölderlin undoubtedly include the fact that Heidegger takes him to be the poet of the Germans, with Heidegger's claim that poetry is 'the primal language of a historical people' (60) making what Philippe Lacoue-Labarthe terms Heidegger's 'politics of poetry' very clear (see Lacoue-Labarthe 2007). In short, Heidegger's thinking of poetry as the counterforce to nihilism is, like all thinking of nihilism, irreducibly political.

Together with Heidegger's claim that Hölderlin is the poet's poet is his insistence on Hölderlin's having named modernity as the 'destitute time' (*dürftige Zeit*) between the withdrawal of the gods and their anticipated return. This time is the subject of the seventh strophe of the elegy 'Bread and Wine' (1801), to which Heidegger turns at the end of 'Hölderlin and the Essence of Poetry'. The strophe reads:

> But friend! we come too late. To be sure, the gods are alive,
>   But up there above our heads in another world.
> Endlessly they are active there and seem to care little
>   If we live, so much do the heavenly ones spare us.
> For a weak vessel cannot always contain them,
>   Only at times can man bear divine fulness.
> So life is a dream about them. But error
>   Helps, like sleep, and distress and night give strength,
> Till heroes enough have grown up in the brazen cradle,
>   Hearts as before have strength like the heavenly ones.
> Then thundering they will come. Meanwhile it often seems to me
>   Better to sleep than to be so companionless,
> Always waiting, and what to do meanwhile and what to say,
>   I know not, and what are poets for in a destitute time?
> But they are, you say, like the wine-god's holy priests,
>   Who passed from land to land in the holy night.
>   (Hölderlin 1951: 93–4; my translation)

Anticipating Nietzsche, Hölderlin here compares the poet of the *dürftige Zeit*, in the absence of the gods, to the 'priests' of Dionysus. The poet keeps alive the thought of the gods in their absence. Modernity thus conceived – in what both Nietzsche and Heidegger, in their different ways, see as its nihilism – is a time that stands, in Heidegger's words, 'in a double lack and a double not: in the no-longer of the gods who have fled and in the not-yet of the god who is coming' (Heidegger 2000a: 64). Heidegger takes Hölderlin to be the poet who 'first determines' this 'new time', and who in so doing heralds its overcoming (64). For Heidegger, then, the answer to Hölderlin's question 'what are poets for in a destitute time' (*wozu Dichter in dürftiger Zeit*) is that the poet not only names this 'new time' as the time of nihilism, but also heralds the overcoming of nihilism through a recollective naming of beings in their being. Crucially, for Heidegger, while writing from within this 'destitute time', the poet's own work is not itself nihilist. Rather, the poet 'holds firm in the Nothingness of this night', stands against it, his work the counterforce to it (65).

In his 1946 essay 'What Are Poets For?', Heidegger returns to the lines from Hölderlin's elegy 'Bread and Wine', and makes it clear that his own position has not been affected either by the Second World War or by the coming to light of the Holocaust. Indeed, Heidegger simply reiterates in this essay that the time to which Hölderlin refers in the poem 'means the era to which we ourselves still belong', this era being 'defined by the god's failure to arrive, by the "default of God"' (Heidegger 1971b: 91). In this default, what can be learned from Hölderlin is that 'Poets are the mortals who, singing earnestly of the wine-god [Dionysus], sense the trace of the fugitive gods, stay on the gods' tracks, and so trace for their kindred mortals the way toward the turning' (94). Heidegger then proceeds to argue that Rainer Maria Rilke fails to equal Hölderlin as a poet who stays on the track of the absent gods, promising the saving resacralization of modernity, since he is essentially Nietzschean in both his thinking and his poeticizing. The Angel of Rilke's *Duino Elegies* (1912–22) is, Heidegger claims, '*metaphysically the same* as the figure of Nietzsche's Zarathustra' (134; Heidegger's emphasis), and thus the consummation of nihilism rather than its overcoming.

The one modern poet who does bear comparison with Hölderlin, in Heidegger's view, is Georg Trakl (1887–1914), whose work he considers in detail in the 1953 essay 'Language in the Poem'. Like Hölderlin's, Trakl's poetry names modernity as the era of decline, darkening, decay, and dissolution, the time when the 'soul' or 'spirit' is a 'stranger on the earth'. And like Hölderlin's, Trakl's poetry is 'historical in the highest sense', by which Heidegger means that his poetry 'sings of the destiny which casts mankind in its still withheld nature – that is to say, saves mankind' (Heidegger 1971a: 196). Trakl's last poem, 'Grodek', is of particular importance to Heidegger because it concerns not simply the 1914–18 war – as is often claimed by commentators on it – but the *dürftige Zeit* of modernity, of which Hölderlin writes, and the 'spirit' (*Geist*) as that from which both destruction and renewal spring. In Alexander Stillmark's translation, the second version of the poem reads:

> At evening the autumn woods resound
> With deadly weapons, the golden plains
> And blue lakes, the sun overhead
> Rolls more darkly on; night embraces
> Dying warriors, the wild lament
> Of their broken mouths.
> Yet silently red clouds, in which a wrathful god lives,
> Gather on willow-ground
> The blood that was shed, moon-coolness;
> All roads flow into black decay.
> Under the golden boughs of night and stars
> Sister's shadow sways through the silent grove,
> To greet the spirits of the heroes, the bleeding heads;
> And softly the dark pipes of autumn sound in the reeds.
> O prouder sorrow! You brazen altars,
> The spirit's ardent flame today is fed by mighty grief,
> The unborn generations.
>
> (Trakl 2001: 127)

The crucial line for Heidegger in this poem is the penultimate: 'The spirit's ardent flame today is fed by mighty grief' (*Die heiße Flamme des Geistes nährt heute ein gewaltiger Schmerz*), its importance lying in its naming of *Geist* as flame. On this line, Heidegger comments: 'Trakl sees spirit not primarily as *pneuma*, something ethereal, but as a flame that inflames, startles, horrifies, and shatters us' (Heidegger 1971a: 179). Spirit thus understood 'has its being in the possibility of *both* gentleness *and* destructiveness' (179; Heidegger's emphasis). This destructiveness is 'active evil', and this active evil is 'the revolt of a terror blazing away in blind delusion, which casts all things into unholy fragmentation and threatens to turn the calm, collected blossoming of gentleness to ashes' (179). It is certainly possible to see Heidegger's reading here as suggesting, without ever stating, that 'active evil' is active nihilism as that which belongs to the essence of spirit and is, for this reason, unavoidable. Taken one step further, this interpretation might even lead one to see Heidegger as proposing an interpretation of the Holocaust as belonging to spirit.

What remains beyond dispute is that Heidegger finds in both Hölderlin and Trakl a poetic naming of modernity as the time of nihilism and a recollection of Being in the era of its oblivion – that is, a recollection of the gods in their absence. In this, poetry would undoubtedly stand as what Nietzsche terms the 'only superior counterforce' to nihilism, resacralizing the world or, at the very least, gesturing towards such a resacralization. Furthermore, just as Nietzsche distinguishes between a Dionysian art of the dithyramb and 'romantic' art, so Heidegger insists on a radical distinction between poetry (*Dichtung*) and literature (*Literatur*). In *What Is Called Thinking?* (1954), for instance, he writes:

> Homer, Sappho, Pindar, Sophocles, are they literature? No! But that is the way they appear to us, and the only way, even when we are engaged in demonstrating by means of literary history that these works of poetry really are not literature.
>
> Literature is what has been literally written down, and copied, with the intent that it be available to a reading public. In that way, literature becomes the object of widely diverging interests, which in turn are once more stimulated by means of literature – through literary criticism and promotion. Now and then, an individual may find his way out of the literature industry, and find his way reflectively and even edifyingly to a poetic work; but that is not enough to secure for poesy the freedom of its natural habitat. (Heidegger 1968: 134)

In 1966, he goes even further, declaring that 'present-day literature' – of which, significantly, he gives no examples – is the antithesis of *Dichtung*. The former is, he claims, 'predominantly destructive' (Heidegger 2003b: 37). For both Nietzsche and Heidegger, then, it is in a very particular form of art – the Dionysian dithyramb for Nietzsche, the *Dichtung* of Hölderlin and Trakl for Heidegger – that the counterforce to the nihilism of modernity is located. In both forms of art, the nihilism of modernity is named, and in both a resacralization of modernity (the return of the gods) is promised.

The role of the concept of nihilism in both the theory and the practice of aesthetic modernism extends far beyond Heidegger, and it is within that larger context that his own thinking of *Dichtung* may be situated. With the popularization of the concept of nihilism in the early decades of the twentieth century through the publication of Nietzsche's *Will to Power*, the charge of nihilism came frequently to be directed at various forms of aesthetic modernism by commentators of all political persuasions, and this continued into the post-Second World War era. On the one hand, nihilism serves here as the key term in a critique that comes from all parts of the political spectrum. On the other hand, it also serves on occasion in the championing of aesthetic modernism. Furthermore, the charge of nihilism is repeatedly countered (either by the writers themselves or by commentators on them) with the claim that what might *appear* to be nihilism is in fact an instance of art as the counterforce to nihilism. This war between the charge of nihilism and the claim of anti-nihilism lies at the very heart of the relation between aesthetic modernism and nihilism, and finds it first major point of focus in the various Dada movements of the years 1916–22.

'There is great destructive, negative work to be done': Dada

Of all the European avant-garde movements of the early decades of the twentieth century, including Futurism in Italy and Russia, Expressionism in Germany, Vorticism in England, and Surrealism in France, it is undoubtedly Dada that has most frequently been charged with nihilism. The Berlin Dadaists' *Dada Almanac* (1920) includes the following extract from an article published in the Boston *Christian Science Monitor*: 'although a correspondent writes to a European paper, "Unfortunately we haven't been able to find out what Dadaism means," he gathers the impression that it stands for "literary nihilism and a complete want of interest in social organization"' (Huelsenbeck 1993: 51). One of the founders of the Surrealist movement, André Breton, accused Dada of 'insolent negation' (quoted in Hopkins 2004: 16), and sought to distinguish Surrealism from Dada in part through a shift of emphasis away from the perceived negativity of the latter. In 1946, Marcel Duchamp, who had been one of the major figures in New York Dada, described the movement as 'a sort of nihilism' and as 'a way to get out of a state of mind – to avoid being influenced by one's immediate environment or by the past: to get away from clichés – to get free' (Duchamp in Lippard 1971: 141). In his retrospective 'Dada at Two Speeds', Marcel Janko, who had been one of the co-founders of Zurich Dada in 1916, distinguishes between the negative and positive 'speeds' in Dada, the former being presented as a necessary phase that was essentially nihilist in nature: 'We had lost confidence in our "culture". Everything had to be demolished. We would begin again after the tabula rasa'; the Dadaists, he claims, 'posed as nihilists declaring art already dead and Dada nothing but a joke' (Janko in Lippard 1971: 36). Janko also remarks upon the reception of Dada, which, he writes, 'was considered everywhere as madness, aggression, derailment, nihilism' (37). In his now-celebrated essay, 'The Work of Art in the Age of Its Mechanical Reproducibility' (1936–9), Walter Benjamin places the emphasis squarely upon the negative in the works of the Dadaists: 'Their poems are "word-salad" containing obscene expressions and every imaginable kind of linguistic refuse. It is not otherwise with their paintings, on which they mounted buttons or train tickets. What they achieved by such means was a ruthless annihilation [*Vernichtung*] of the aura in every object they produced, which they branded as a reproduction through the very means of production' (Benjamin 2003: 267).

A number of influential post-war commentators have taken a similar line. In his *Theory of the Avant-Garde* (1962), Renato Poggioli argues that while there are nihilist tendencies in both Italian and Russian Futurism, as well as in English Vorticism, 'it was perhaps only in dadaism that the nihilistic tendency functioned as the primary, even solitary, psychic condition; there it took the form of an intransigent puerility, an extreme infantilism'; indeed, 'the nihilistic tendency in its pure state demonstrably attained its most intense and varied expression in dadaism', the Dada manifestos announcing 'a totally nihilistic attitude' (Poggioli 1968: 62–3). The nihilism of Dada is, he argues, not merely aesthetic, but 'radical and totalitarian, integral and metaphysical' (62–3). That said, Poggioli restricts Dada to the realm of the aesthetic by claiming that the nihilism not only of Dada but also of the historical avant-garde more generally should be understood as a response to the 'debasement' of art in the West in the modern era:

> Doubtless the nihilistic posture represents the point of extreme tension reached by antagonism toward the public and tradition; doubtless its true significance is a revolt of the modern artist against the spiritual and social ambiance in which he is destined to be born and to grow and to die. The motivations for this revolt appear simultaneously under the different guises of reaction and escape: reaction against the modern debasement of art in mass culture and popular art; escape into a world very remote from that of the dominant cultural reality, from vulgar and common art, by dissolving art and culture into a new and paradoxical nirvana. (64)

In *The Revolution of Everyday Life* (1967), the Belgian theoretician of, and prime mover in, the 1960s Situationist movement, Raoul

Vaneigem, returns to Nietzsche's distinction between active and passive nihilism, arguing that 'The active nihilist does not simply watch things fall apart. He criticizes the causes of disintegration by speeding up the process' (Vaneigem 1983: 136). According to Vaneigem, Dada exhibits just such an active nihilism, this movement being the one among the various historical avant-gardes in which the awareness of cultural decay 'reached its most explosive expression', as did the contempt for 'art and bourgeois values' (137–8). For this reason, Vaneigem sees Dada as a key forerunner of the Situationists.

More recently, the Dada scholar David Hopkins has argued that it is possible to distinguish between Swiss, German, and French Dada in terms of the extent to which they were nihilist in their orientation. Whereas the Dada of Hugo Ball, Tristan Tzara and others at the Cabaret Voltaire in Zurich, in 1916, was an outright attack on bourgeois morality and aesthetics, it was, Hopkins argues, not nihilistic in the way that the French Dada of Francis Picabia was. Indeed, according to Hopkins, Picabia was Dada's 'self-appointed ambassador of nihilism'. That said, Hopkins believes that even Zurich Dada became 'nihilistic' in its 'final phase' (Hopkins 2004: 14).

A similarly nuanced view of the avant-garde and of modernism more generally is taken by Richard Sheppard, who distinguishes between nine responses to the 'perceived crisis of modernity' in aesthetic modernism. These include the 'ultrapessimistic or nihilistic response', to be found in the works of Georg Trakl and Franz Kafka. According to Sheppard, Dada is *not* among such nihilistic responses, the reason for this being that its principal figures belong with those modernists 'who accept that modernity is in crisis but refuse to succumb to despair, mindless irrationalism, or nostalgia; who affirm modernity but reject modernolatry and the desire for closure' (Sheppard 2000: 86). Like Hopkins, however, Sheppard does find nihilist tendencies in Dada. He argues, for instance, that after 1918 Tristan Tzara 'fell prey to the dire sense of apathy that marks his letters to Breton and Picabia of 1919 and reverted to an aggressive nihilism, untempered by any concern with the spirit, during the last year of his involvement with Dada in Zurich and throughout his involvement with Dada in Paris (late 1918–late summer 1922)' (187).

It is not hard to see why so many commentators have characterized Dada as nihilist. From the outset, its practitioner-theorists were placing the idea of 'nothingness' at the heart of the Dada response to what they took to be the crisis of modernity and its values, above all the values of reason, progress, and (as Peter Bürger argues in *The Theory of the Avant-Garde*, 1974) art in its autonomy from life. In *Declaration* (*Erklärung*), a short text delivered at the Cabaret Voltaire, Zurich, in the spring of 1916, Richard Huelsenbeck says of the word 'Dada' that it 'signifies nothing' (*bedeutet nichts*), although he then nuances this statement with the claim that it is 'the significant/signifying nothing' (*das bedeutende Nichts*). The importance of this *Nichts* becomes clear when he elaborates on Dada's aims: 'We wish to change the world with nothing, we wish to change poetry and painting with nothing, and we wish to bring the war to an end with nothing' (in Riha and Schäfer 1994: 33; my translation). The Dada 'nothing' thus conceived – as radically transformative – is squarely in line with Nietzsche's conception of active nihilism.

A similar position is taken by Tristan Tzara in his 'Dada Manifesto', first published in *Dada* 3 in December 1918. Like Huelsenbeck, Tzara insists that 'DADA DOES NOT MEAN ANYTHING' (Tzara 1992: 4). He then declares that the Dadaists are 'like a raging wind that rips up the clothes of clouds and prayers, we are preparing the great spectacle of disaster, conflagration and decomposition', that 'There is no Truth', that 'Logic is always false', and that 'Every man must shout: there is great destructive, negative work to be done. To sweep, to clean' (8–12). The targets for all this 'negative work' include the family, logic, memory, and the future. Two years later, in his 'Dada Manifesto on Feeble Love and Bitter Love', Tzara turns this negative energy back against Dada itself, declaring that 'the real dadas are against DADA' (38). This position is elaborated in the 1922 *Lecture on Dada*, in which Tzara states that 'Dada is nothing' and that 'I parted from Dada and from myself the moment I realized the true implication of *nothing*' (107; Tzara's emphasis).

The apparently nihilistic elements in Dada arguably find their most forceful expression in the declarations of Francis Picabia, whose 'DADA Manifesto', published in *391 Paris* 12 (March 1920), ends:

> As for Dada it means nothing, nothing, nothing. It makes the public say, 'We understand nothing, nothing, nothing.'
> The Dadaists are nothing, nothing, nothing, they will certainly succeed in nothing, nothing, nothing.
> Francis PICABIA
>   who knows nothing, nothing, nothing.
>
> (Picabia in Ades 2006: 125)

In another Dada manifesto, this one published in *Littérature* 13 (May 1920), Picabia fleshes out his position as follows:

> No more painters, no more writers, no more musicians, no more sculptors, no more religions, no more republicans, no more royalists, no more imperialists, no more anarchists, no more socialists, no more Bolsheviks, no more politicians, no more proletarians, no more democrats, no more bourgeois, no more aristocrats, no more armies, no more police, no more fatherlands, enough of all these imbecilities, no more anything, no more anything, *nothing, nothing, nothing, nothing.*
>
> We hope something new will come from this, being exactly what we no longer want, determinedly less putrid, less selfish, less materialistic, less obtuse, less immensely *grotesque*. (181; Picabia's emphasis)

And again, in his 'Cannibal Dada Manifesto', published in *The Dada Almanac* (1920), Picabia places the emphasis relentlessly on the idea of the nothing:

> DADA, as for it, it smells of nothing, it is nothing, nothing, nothing.
>   It is like your hopes: nothing
>   like your heaven: nothing
>   like your idols: nothing
>   like your politicians: nothing
>   like your heroes: nothing
>   like your artists: nothing
>   like your religions: nothing.
>   (Picabia in Huelsenbeck 1993: 56)

This recurrent insistence upon negativity and the nothing is, however, very far from being the whole story of Dada. To be sure, the Dadaists wished to express in the strongest terms their rejection of an entire society, its culture, and its values, all of which they considered to have been utterly discredited by the 1914–18 war. This rejection found a crucial forefather in Nietzsche, who is arguably the overriding influence on the Dadaists. For that very reason, however, Dada could not be an exclusively negative movement. Just as Nietzsche's critique of modernity as nihilist entails a thinking of the overcoming of nihilism through an affirmation of 'life', so many of the Dadaists sought to accomplish a movement from negation to affirmation, with the former being a phase or, as

Marcel Janko puts it, one of Dada's two 'speeds' (*vitesses*). Indeed, members of the Dada movement often sought to distinguish their positions from nihilism. In the 1920 *Dada Almanac*, for instance, Huelsenbeck asserts that 'Whoever turns "freedom", or "relativity", including the insight that the contours of everything shift, that nothing is stable, into a "firm creed" is just another ideologue, like the nihilists who are almost always the most incredible, narrow-minded dogmatists. Dada is far removed from all that' (Huelsenbeck 1993: 11).

The insistence upon both critique and affirmation is nowhere more evident than in the writings of Hugo Ball (1886–1927), one of the founders of Zurich Dada. Ball had abandoned a doctoral thesis on Nietzsche a few years before the coming into being of Dada in 1916, and Nietzsche is a major presence in his published diary, *Flight Out of Time* (1927). It is doubtless in no small part on account of his interest in Nietzsche's thought that the concept of nihilism looms so large in Ball's work. On occasion, Ball's diary reveals just how close even he comes to conceiving of Dada as nihilist. The 12 June 1916 entry, for instance, includes the assertion that 'What we call dada is a farce of nothingness [*ein Narrenspiel aus dem Nichts*] in which all higher questions are involved: a gladiator's gesture, a play with shabby leftovers, the death warrant of posturing morality and abundance' (Ball 1996: 65). Taken as a whole, however, Ball's diary shows him to be committed to distinguishing his own position from that of the nineteenth-century Russian nihilists. As a diary entry of 25 November 1914 demonstrates, Ball considers nihilism to be too conservative rather than too negative. Russian nihilism clings to reason, which is precisely what Dada will not do:

> The nihilists base their ideas on reason (their own). But we must break with the system of reason, because a higher reason exists. The word 'nihilist', by the way, means less than it says. It means: one cannot rely on anything, one must break with everything. It *appears* to mean: nothing can remain in existence. They want to have schools, machines, a rational economy, and everything that Russia still lacks but that we in the West have much too much of. (11–12; Ball's emphasis)

Three years later, after his break from Dada, Ball continues to reflect on Russian nihilism, at a time when he was working on a book about the Russian anarchist Mikhail Bakunin, whose work he believes anticipates 'all of Nietzsche' (136). In a diary entry of 29 August 1917, for instance, Ball writes:

> Nihilism, as Pisarev and Zajzev preached it, was the protest of groups who lived in tolerable material and social conditions but suffered under the pressure of conventional customs and ideas. They sought the freedom of the individual and fought all intellectual and moral chains. (We have had more than enough of that. Imitation of it could be only an anachronism. While people are drawing practical consequences from our obsolete theories, we are already preparing ourselves for ideological conversion.)
>
> Nihilism in Russia, just as in the West, smoothed the way for anarchism. (129)

In the years following Dada's explosion onto the European cultural scene, Ball set out to distinguish his own position from Nietzsche's, above all regarding the relation between art and religion. As he rightly observes, Nietzsche came to see art as that which stood directly opposed to Christianity, since, Nietzsche argued, art was not tied to the value of truth. This is the position articulated in the third essay in *On the Genealogy of Morals* (1887), where Nietzsche claims that in art 'the *lie* is sanctified and the *will to deception* has a good conscience' (Nietzsche 1989: 153; Nietzsche's emphasis). In a diary entry of 4 December 1917, Ball counters this view with the claim that, in fact, 'Art is much closer to religion than science is. To me it is an incomprehensible antithesis when Nietzsche sets art against a union of religion and science' (Ball 1996: 94). To Nietzsche's claim in the preface to the 1886 edition of *The Birth of Tragedy* that Christianity negates aesthetic values, Ball responds in a diary entry of 18 December 1917: 'Is that right? Franz von Baader […] does the opposite and characterizes religion, and thus Christianity, as the higher poetic art' (95). Ball seeks to reintroduce the concept of truth into art by way of the notion of 'fictive truths', asserting that a poem 'is true without one's being able to prove it in reality' (95). Furthermore, art is not to be understood as purely negative in essence. Rather, its negative moment clears the ground for an affirmation of new values: 'Art enlarges the world by negating aspects that were known and in operation up to now, and putting new ones in their place. That is the power of modern aesthetics; one cannot be an artist and believe in history' (22 March 1917; 101).

In Ball's own brand of Dada poetry – what he termed his 'poems without words' (*Verse ohne Worte*) or 'sound poems' (*Lautgedichte*) – there is just such a play of negation and affirmation. This 'new genre of poems' (70) includes 'Karawane', which was recited by Ball at the Cabaret Voltaire on 23 June 1916:

 jolifanto bambla ô falli bambla
 grossiga m'pfa habla horem
 égiga goramen
 higo bloiko russula huju
 hollaka hollala
 anlogo bung
 blago bung
 blago bung
 bosso fataka
 ü üü ü
 schampa wulla wussa ólobo
 hej tatta gôrem
 eschige zunbada
 wulubu ssubudu uluw ssubudu
 tumba ba- umf
 kusagauma
 ba – umf
            (in Schaffner 2007: 73)

On the one hand, this poem clearly aims at the complete negation of what, in a diary entry of 24 June 1916, Ball describes as 'the language that journalism has abused and corrupted' (Ball 1996: 71). On the other hand, this negation leads neither to silence nor to pure meaninglessness, but rather to a new relation of word and world. In place of a dead language, new words are to be created to capture the essence of things. According to Ball, in poetry there should be a return to 'the innermost alchemy of the word', a 'newly invented language for our own use' (71). It might seem that such a conception of language goes against the idea of language as a means of communication, since it appears to reject any conventional or shared agreement on the relationship between signifier and signified of the kind emphasized by the linguist Ferdinand de Saussure in his 1907–11 lectures at the University of Geneva, first published in 1916 as *Course in General Linguistics*. Certainly, Ball wishes to establish new connections between these two sides of the linguistic sign, and to produce effects that are unique for each listener. As he puts it in his *Dada Manifesto* of 14 July 1916, Ball strives to free himself from 'other people's inventions' and to produce poetry that is absolutely singular: 'I want my own stuff, my own rhythm, and vowels and consonants too, matching the rhythm and all my own' (221). But this aim is clearly distinct from that of producing a text that is simply non-signifying, meaningless, or, as Huelsenbeck puts it, 'signifies nothing' (*bedeutet nichts*). Rather, each listener or reader will necessarily take something different from the poem, even if many of the signifiers might seem to bear a striking resemblance to others from the 'abused and corrupted' language that has hitherto been the language not only of journalism but also of art. For instance, the word 'jolifanto' in the poem 'Karawane' appears to suggest 'elephant', while 'grossiga' appears to derive from the German word for 'large' (*groß*), and the poem as a whole might even be taken to evoke the sound and movement of an elephant.

An intimate relation between negation and affirmation is also evident in Ball's important lecture on the art of the Russian abstract painter Wassily Kandinsky, delivered in Zurich on 7 April 1917. The lecture begins with a characterization of 'The Age' that is clearly Nietzschean in its inspiration. Although the term 'nihilism' is not used by Ball in this lecture, his vision of modernity matches perfectly Nietzsche's conception of the modern age as the consummation of nihilism:

> God is dead. A world disintegrated. [...] An epoch disintegrates. A thousand-year-old culture disintegrates. There are no columns and supports, no foundations any more – they have all been blown up. [...] Above is below, below is above. The transvaluation of values came to pass. Christianity was struck down. The principles of logic, of centrality, unity, and reason were unmasked as postulates of a power-craving theology. The meaning of the world disappeared. The purpose of the world – its reference to a supreme being who keeps the world together – disappeared. (223)

Ball goes on to declare, contra an entire philosophical tradition commencing with Protagoras, that the human being is no longer the measure of all things, that humanity has become merely 'a particle of nature', that one can no longer appeal to reason, that the machine has replaced the individual, and that the life of the modern artist is a perpetual 'struggle against madness' (224–5). On the one hand, this struggle must take the form of a negation directed not only against society, but also against art: 'The artists of this age turn against themselves and against art.' On the other hand, true artists are 'forerunners, prophets of a new age' (225). The artist's task is not simply the negation of existing values, but, as Nietzsche claimed, their 'transvaluation' (*Umwerthung*). As for Kandinsky, his paintings are 'an exit from the confusions, the defeats, and the doubts of the age. They are a liberation from the thousand years that are now on the point of disintegrating' (226). In short, art is clearly presented by Ball here as Nietzsche's 'superior counterforce' to nihilism. Although, as we have seen, he comes to take his distance from Nietzsche, and above all from the Nietzsche who announces the 'death of God' – Ball even declaring in a 17 April 1921 diary entry that 'Dionysius the Areopagite is the refutation of Nietzsche in advance' (201) – he does not abandon his faith in art as such a counterforce, and on that account he remains tied to what is undoubtedly among the most important elements in the Nietzschean heritage.

According to Alain Badiou, the twentieth century conceived of itself 'simultaneously as end, exhaustion, decadence *and* as absolute commencement. Part of the century's problem is the conjunction of these two convictions. In other words, the century conceived of itself as nihilism, but equally as Dionysian affirmation' (Badiou 2007: 31; Badiou's emphasis). The problem of this conjunction is nowhere more evident, nowhere more pressing, than in Dada, where a distinctly Nietzschean emphasis upon the decadence of modernity, the need to negate all the existing values of that modernity, and the urgent call for the introduction of new values are all present. It is less obviously the case, however, in the work of the writer who, arguably more than any other, is the paradigmatic figure in the literature of 'high' modernism: Franz Kafka. It is to a consideration of Kafka and those who follow in his wake that we may now turn.

4

KAFKA AND AFTER

Aesthetic modernism as 'new nihilism'

As we have seen in the previous chapter, while the characterization of aesthetic modernism as a form of nihilism is usually a means of critiquing it, this is certainly not always the case. Those Dadaists such as Marcel Duchamp who characterized Dada as a nihilist movement in the arts were championing what they saw as a necessary act of negation, a destruction of values, forms, and relations that they took to be decadent, inhibiting, or reactionary. Such a championing of aesthetic modernism as a form of nihilism is far from being limited to avant-garde movements of the immediate post-First World War years, as is evidenced by the publication of an article entitled 'The New Nihilism' in the May 1927 issue of the Paris-based literary magazine *transition*, in which the work of many of the now-canonical 'high' modernists appeared in the late 1920s and 1930s, including Franz Kafka, James Joyce, Gertrude Stein, Ernest Hemingway, and Samuel Beckett.

In 'The New Nihilism', *transition*'s co-editor, Elliot Paul, argues that while, as a result of the 1914–18 war, the 'old values had become meaningless' (Paul 1927: 164), in the immediate aftermath of that war no new literary movement had arisen that took adequate account of this fact: 'Great books were not forthcoming. Sound conclusions did not spring up from the wreckage which littered all Europe' (165). The great literature of the pre-war period was, he claims, shaped by a humanist ideology, by the belief that 'human brotherhood could be realized by the awakening in each man and woman of their unselfish and kindly instincts' (165). According to Paul, this humanism found its most comprehensive and profound expression in the novels of Fyodor Dostoevsky during the 1860s and 1870s. The 1914–18 war had destroyed such beliefs, however, and now, a decade later, a new, post-humanist literature was finally emerging. This new literature was 'completely dehumanized' and knew 'neither morals nor compassion' (165–6). Paul insists, however, that this post-humanism is not to be confused with Nietzscheanism. It is amoralist rather than immoralist in nature, and the 'new hero' is no 'superman' in the Nietzschean sense, since he 'neither feels nor shows superiority, only an utter amorality and a clear head which finds futility everywhere and accepts it as a natural law' (166). Paul goes on to identify the purest form of this new, post-humanist literary movement as occurring in France, and, above all, in a work entitled *The Young European* by Pierre Drieu La Rochelle (1893–1945). To give his anglophone readers a taste of this 'new nihilism', Paul included his own English translation of the first chapter of Drieu La Rochelle's work in the issue of *transition* in which his essay appeared.

In *The Young European*, Paul finds a literature that is antithetical to Dostoevsky's. Whereas Dostoevsky sought to offer a compelling critique of Russian nihilism through the literary, this work of 'new nihilism' is, he argues, 'a frank and lucid and convincing statement of a world intellectual tendency which is in its ascendancy. It renounces Christ and Nietzsche as if both were schoolboys. No illusions as to the revival of Europe's past greatness can live in its atmosphere, and America's noise and activity, the glitter of the dollar and the whirr of the machine, are stripped of constructive value' (166–7). Paul praises *The Young European* as a work of 'perfect inhumanity' that 'goes way beyond the Russian Nihilism of Turgenev's time' (167–8). Paul's own conception of this 'new nihilism' nonetheless bears a close relation to nineteenth-century Russian nihilism in that he sees the new movement as breaking radically with existing institutions. For Russian nihilism, these institutions were principally political; for Paul's 'new nihilism' they are also aesthetic, moral, and metaphysical.

Although Paul does not register the fact, Drieu La Rochelle's conception of modern humanity in *The Young European* bears some striking similarities to that of the German writer Ernst Jünger in his account of his experiences on the Western Front, *In Storms of Steel: From the Diary of a Storm Troop Leader* (1920). The war marks 'the end of vanity' and the beginning of a new experience of man as 'workman'

(Drieu La Rochelle 1927: 10). It discloses modern man as a violent being best suited to destruction: 'The violence of men. They are born only for war, as women are made to have children. All the rest is a tardy detail of the imagination which has already shot its bolt. [...] Man need never have left the forest: he is a degenerate, nostalgic animal' (11). The war enables 'the rupture and the dissolution of all matter, of all ties' (12). The 'decadence' of the liberal-bourgeois epoch is countered by a liberating, if murderous, violence: 'It is necessary to have killed with the hands to understand life. The only life of which men are capable, I tell you again, is the spilling of blood: murder and coitus. All the rest is but the fag end of the course, decadence' (17). The work's opening chapter clearly foreshadows Drieu La Rochelle's later commitment to fascism. Neither American capitalism nor Russian communism offers a genuine alternative to the decadence of modernity:

> The Americans were nothing but worse Europeans who had changed continents to play more at ease their game of brutes captivated by the abstract. [...] I had deceived myself. The Russian revolution was not at all as I had believed. Those Jews thought only of making themselves Americans, only, like the Germans of 1914, they went about it awkwardly. (16–17)

Paul's championing of a 'new nihilism' is anticipated by *transition*'s other founding editor, Eugene Jolas, who, in an article on the German poet Gottfried Benn (1886–1956) published in *transition* 5 (1927), characterizes modernity as nihilist in the following manner: 'The tragic feeling of standing isolated in his cosmos is a fate the sensitive man of today cannot escape. In spite of the progress of machines, this ever present antinomy between his wishes and his phenomenal world produces in him a negative attitude, a nihilism that is a paralysis' (Jolas 1927: 146). According to Jolas, Benn is a 'courageous' poet not because he offers a countering affirmation to this nihilism, but because he fights the nihilism of modernity with an *aesthetic* nihilism: 'He brutally and consciously works for the disintegration of the universe he knows' (149).

As for Benn's own view on nihilism, the title essay in his collection *After Nihilism* (1932) makes it clear that he follows Nietzsche not only in conceiving of modernity as nihilist, but also in privileging aesthetic values as those through which this nihilism can be overcome. According to Benn, the nihilism of modernity originates in the mid-nineteenth-century reconception of the world as an intelligible mechanism (that is, as an object of physics) and in the Darwinian reconception of man in relation to other animal life, at the heart of which lay the principle of the survival of the fittest. Benn defines the consequent nihilism as 'the dissolution of all old ties, the destruction of substance, the levelling of all values' (Benn 1970: 100). As Michael Hamburger observes in 'Art and Nihilism' (1954), Benn 'believes that Nihilism is the inevitable frame of mind of all those Europeans of the present age who have the courage to think' (Hamburger 1954: 52). This view is certainly borne out by Benn's claim in his essay 'Art and the Third Reich' (written in 1941; published in 1949) that 'For centuries all the great men of the white race had felt only the one inner task of concealing their own nihilism' (Benn 1976: 93). Among the major figures to have responded in this way to nihilism, Benn includes Dürer, Tolstoy, Goethe, and Balzac. For each of these artists, the 'inner creative substance' is 'the abysmal, the void, the cold, the inhuman' (93). As for Nietzsche, he is the one 'who remained naïve the longest', since it is not in what, in another essay of the same period, Benn terms the 'evolutionary optimism' of his *Zarathustra*, but only in his last work, *Ecce Homo*, that Nietzsche joins the ranks of those who have engaged authentically with the experience of nihilism.

Benn's principal objection to Nietzsche concerns the latter's conviction that the overcoming of nihilism conceived as a levelling of all values can be achieved by the overman (*Übermensch*). Benn dismisses the concept of the overman as Darwinian or '*biologically positive*' (Benn 1970: 102; Benn's emphasis). According to Benn, the overcoming of nihilism will be accomplished only when a 'new *ethical* reality' is produced through a commitment to 'aesthetic values', and more precisely to 'the educated absoluteness of form', to 'linear purity' and 'stylistic immaculateness' (104). As what in his essay 'Problems of the Lyric' (1951) he terms 'artistics' (*Artistik*), Benn conceives of aesthetic practice as the sole counterforce to the prevailing nihilism: '*Artistik* is the attempt of Art to experience itself as a meaning within the general decay of all meaning, and to form a new style out of this experience; it is the attempt of Art to oppose the general nihilism of values with a new kind of transcendence, the transcendence of creative pleasure' (Benn 2001: 14; quoted in Hamburger 1954: 54). That said, even in the post-war era in which he formulates this response to nihilism,

Benn ties the commitment to the aesthetic in the struggle against a loss of values not only to the concept of the nation, but to the *German* nation. As he puts it in 'After Nihilism', his concern is 'an entirely new moral and metaphysic of form for Germany', with the 'law of form' as a 'national obligation' (Benn 1970: 103–4).

Crucially, however, Benn insists that 'artistics' or 'absolute' art – that is, art as pure style – is in fact another form of nihilism. This is made clear when Benn claims that 'Nihilism as the negation of history, actuality, affirmation of life, is a great quality, but as the negation of reality itself it means a diminution of the ego' (quoted in Hamburger 1954: 57). By distinguishing between two forms of nihilism, Benn follows Nietzsche in his distinction between active and reactive nihilism, the former being a nihilism that is required if the overcoming of nihilism is to be achieved. That said, by 1954 Benn goes so far as to dismiss the term 'nihilism' altogether, arguing in a radio broadcast entitled 'Nihilism or Positivism? On the Position of Modern Man' that 'We may as well delete the word "nihilism". For the last two decades it has lost virtually all meaning. [...] Modern man does not think nihilistically; he puts order into his thoughts and thus creates a basis for his existence' (Benn 1976: 209). As in the pre-Second World War era, so here too Benn accords a privilege to art and to the 'creative man', who, he claims, 'even if personally and privately afflicted with the deepest pessimism, would rise from the abyss by the mere fact that he works. The accomplished work itself is a denial of decay and doom' (210). As we shall see, Jolas is not alone in characterizing as nihilist Benn's own pre-war art, and especially the collections *Morgue and Other Poems* (1912) and *Flesh* (1917), for this idea is taken up in the later 1930s by another major theorist of modernism, Walter Benjamin.

Although Elliot Paul undoubtedly champions a new movement that was soon to lead straight to the most extreme form of right-wing politics, his article on the 'new nihilism' elicited a polemical response from one of the major modernist writers of the political Right in Britain, Wyndham Lewis, who as early as 1931 was to publish a book on Hitler that was largely supportive of Nazi policy. In the second part of *The Diabolical Principle* (first published in the third and, as it happened, final issue of his journal *The Enemy: A Review of Art and Literature*, in 1929), Lewis does not limit himself to an attack on Paul's argument regarding the 'new nihilism'. Rather, he engages in a polemical attack on the entire modernist movement, including writers such as Gertrude Stein and James Joyce. According to Lewis, this movement is in fact nothing other than 'romantic nihilism' (Lewis 1994: 40), a label that he also applies to Nietzsche, 'nihilism' here being given an unambiguously pejorative sense. Disregarding Paul's claims to the contrary, Lewis argues that Drieu La Rochelle's 'new nihilism' is not only thoroughly Nietzschean in inspiration but also precisely the kind of nihilism subjected to such harsh critique by Dostoevsky in the 1870s:

> *The Possessed* of Dostoieffski describes exactly the same sort of 'nihilists' as Paul is concerned to advertise. The strictly 'inhuman' or rather anti-human vindictiveness that makes possible the massacres of the various contemporary Revolutions, is a 'nihilism': it has to its credit a holocaust. But substantially it is the same as the demented doctrine of universal destruction which Dostoieffski despairingly observed, and put on record with such clairvoyance. (32)

Lewis traces what he takes to be a nihilist literary tradition originating in mid-nineteenth-century French literature, in Baudelaire's *Flowers of Evil* (1857) and Lautréamont's *Songs of Maldoror* (1868–9), the latter defying the laws of traditional intellectual history by being 'a kind of happy mixture' of the Marquis de Sade and Nietzsche (34). In Baudelaire, Lautréamont, Rimbaud, and Nietzsche, Lewis finds the same 'diabolical principle' at work as is invoked by the principal editor of *transition*, Eugene Jolas. Lautréamont's work, for instance, is characterized according to Lewis by 'a spirit of demented hatred of other men, and an obsessional attachment to apocalyptic images of horror and destruction', and this 'destructive hatred' is repeated in the editorial stance of *transition* (39–40). Crucially, Lewis sees this not-so-new 'new nihilism' as rooted in the political situation in post-war Europe: 'The tenor of the philosophy of *Transition* is then a romantic nihilism, and the springs of its action are to be sought not in any specific doctrine of art, but, as its name suggests, in the political chaos of this time, and in a particular attitude towards that chaos' (40). Drieu La Rochelle's work, together with his later commitment to fascism, certainly supports such a conclusion.

Over two decades and a second world war later, in a chapter entitled 'Twentieth Century Nihilism' in *The Writer and the Absolute* (1952), Lewis returns to the question of nihilism, although this time he directs the charge not at the first wave of modernists but at

French existentialism and above all at the work of Jean-Paul Sartre and Albert Camus. Here, Lewis locates the nihilism of modernism in its subjectivism; that is, a negation of the external world originating in the phenomenology of Edmund Husserl, and, in particular, in its call for a phenomenological bracketing (*epochē*) of anything beyond consciousness: 'So – having cut himself off from the phenomenal world outside – in this empty shell our Existentialist flings himself on the floor and contemplates this echoless vacuity' (Lewis 1952: 126). Among the artists now seen by Lewis as nihilist on account of their subjectivism, or their detachment from a shared object world, are not only the major French existentialist writers of the post-Second World War period such as Sartre and Camus, but also André Gide and André Malraux, the philosopher Martin Heidegger, and the painters Pablo Picasso and Paul Klee.

In an irony that is characteristic of the history of the deployment of the concept of nihilism after Nietzsche, the politically right-wing Lewis defines nihilism here in a manner that is closely akin to that of the Hungarian Marxist critic Georg Lukács, to whose critique of modernism as nihilism we shall turn shortly. Before doing so, however, it is first necessary to distinguish Lewis's position on the nihilism of that 'high' modernism championed by the editors of *transition* from the argument that modernist literature often thematizes nihilism. An important example of the latter is Ernest Hemingway's short story 'A Clean, Well-Lighted Place', first published in 1926 and later included in the collection *Winner Take Nothing* (1933). In this story, the thoughts of a café waiter are clearly intended to suggest a nihilist outlook:

> What did he fear? It was not fear or dread. It was a nothing that he knew too well. It was all a nothing and a man was nothing too. It was only that and light was all it needed and a certain cleanness and order. Some lived in it and never felt it but he knew it was all nada y pues nada y nada y pues nada. Our nada who art in nada, nada be thy name thy kingdom nada thy will be nada in nada as it is in nada. Give us this nada our daily nada and nada us our nada as we nada our nadas and nada us not into nada but deliver us from nada; pues nada. Hail nothing full of nothing, nothing is with thee. (Hemingway 1934: 31–2)

Hemingway's short story certainly explores the idea of nihilism, but that does not necessarily make the author or his work as a whole nihilist. All depends on the way in which the nihilism is framed, both within the specific work and within the writer's oeuvre more generally. Indeed, as Norman Podhoretz rightly observes in a 1958 article with the same title as Elliot Paul's of two decades earlier – 'The New Nihilism' – the idea of a 'loss of values [...] informed almost every poem and novel of the modernist movement' (Podhoretz 1958: 576). Writers for whom this would be the case include, to name only some of the most significant: Gottfried Benn, Franz Kafka, Robert Musil, and Thomas Mann in German-speaking countries; Antonin Artaud, Louis-Ferdinand Céline, and André Malraux in France; and T. S. Eliot, D. H. Lawrence, Wyndham Lewis, and Henry Miller in the anglophone world. For instance, in his uncompleted *magnum opus*, *The Man Without Qualities* (1930–42), Musil has his protagonist, Ulrich, conclude that modern man – as the 'man without qualities' – is 'a nihilist' (Musil 1952: 1150). And in *Women in Love* (1920), Lawrence repeatedly characterizes attitudes as nihilistic. In one of the many discussions on the question of love in the novel, Birkin dismisses his lover Ursula Brangwen's belief that 'love includes everything' as 'Sentimental cant', and declares: 'You want the state of chaos, that's all. It is ultimate nihilism, this freedom-in-love business, this freedom which is love and love which is freedom' (Lawrence 1987: 152). And of Winifred Crich, the narrator asserts that her will is 'anarchistic, almost nihilistic', and that she is an individual who lives 'without attachment or responsibility beyond the moment, who in her every motion snapped the threads of serious relationship, with blithe, free hands, really nihilistic, because never troubled' (220). Podhoretz goes on to argue that this theme of a loss of values returns in post-Second World War literature, 'only this time it seems to be taking the form of a recognition that in losing our taste for ideology we have also lost our capacity for passion' (Podhoretz 1958: 580).

In some cases, this 'new nihilism' is presented from a critical vantage-point, as it is, according to Podhoretz, in Albert Camus' 1942 novel *The Stranger*, the narrator of which, Meursault, displays what Podhoretz describes as 'pathological apathy' (584). Whereas Wyndham Lewis sees Camus as nihilist, Podhoretz adjudges him, both in *The Stranger* and in the stories collected in *Exile and the Kingdom* (1957), as striking a 'very delicate balance [...] between identification with the nihilists he writes about and detachment from them' (584–5). On the other hand, Podhoretz finds a complete identification with nihilism in the novels of another French writer who came to prominence in the post-war decades, Nathalie Sarraute, whose work is usually placed by commentators in the category of the 'new novel' (*nouveau roman*), alongside that of writers such as Alain Robbe-Grillet and Claude Simon. In Sarraute's novels, Podhoretz detects 'a total submission to the meaningless[ness] of existence' and a 'coolly presented picture of nothingness' (585). This tendency reaches its most extreme form in Sarraute's novel *Portrait of a Man Unknown* (1948), described by Jean-Paul Sartre in his introduction to the first edition as an 'anti-novel'. For what is arguably the paradigmatic case of a writer whose work has generated a critical literature repeatedly centred on the question of whether or not it is nihilist, however, one has to return to the period of 'high' modernism between the wars.

'We are nihilistic thoughts': Kafka and the negative

Just as Nietzsche lies at the heart of philosophical modernism, so Franz Kafka (1883–1924) is arguably the paradigmatic figure in aesthetic modernism, being the writer who most fully embodies the 'high' modernist project within the sphere of the literary. It is no coincidence that, more than any other major writer of the modernist period, Kafka should also have been the one around whose work the question of nihilism looms largest. The novelist Franz Werfel, an acquaintance of Kafka's in Prague, said of him: 'I would love Kafka much more, if he were not so nihilistic' (quoted in Glicksberg 1975: 125). The charge of nihilism directed at Kafka's works finds its most forceful advocate, however, in the Marxist critic Georg Lukács. According to Lukács, aesthetic modernism in its entirety is 'based on nihilism', with Kafka's oeuvre being the most fully realized form of this nihilism (Lukács 1963: 81). Lukács reads Kafka's posthumously published novels *The Trial* (1925) and *The Castle* (1926) as 'nihilistic allegories' in which nothingness (*das Nichts*) replaces God in a version of Nietzsche's conception of the 'death of God' informed by Jewish mysticism:

> The supreme judges in *The Trial*, the castle administration in *The Castle*, represent transcendence in Kafka's allegories: the transcendence of Nothingness. [...] If there is a God here, it can only be the God of religious atheism: *atheos absconditus*. [...] The hidden, non-existent God of Kafka's world derives his spectral character from the fact that his own non-existence is the ground of all existence; and the portrayed reality,

uncannily accurate as it is, is spectral in the shadow of that dependence. The only purpose of transcendence – the intangible *nichtendes Nichts* – is to reveal the *facies hippocratica* of the world. (44–5)

In the course of his critique of Kafka's oeuvre as nihilist, Lukács refers to Walter Benjamin's influential 1934 essay on Kafka, in which Benjamin cites the writer's remark to his friend and biographer Max Brod: 'We are nihilistic thoughts, suicidal thoughts, that come into God's head' (quoted in Benjamin 1999b: 798; cf. Lukács 1963: 43). Neither in his 1934 essay, nor in his earlier (1931) essay on Kafka, however, does Benjamin seek to demonstrate that Kafka's works are nihilist. Indeed, Benjamin's correspondence with the scholar of Jewish mysticism Gershom Scholem during the 1930s reveals that it is Scholem rather than Benjamin who finds a form of religious nihilism in Kafka.

In a letter to Benjamin dated 9 July 1934, Scholem encloses what he describes as a 'theological didactic poem' entitled 'With a Copy of Kafka's *The Trial*' (Scholem 1989: 122). This poem includes the following lines:

> The great deceit of the world
> Is now consummated.
> Give then, Lord, that he may wake
> Who was struck through by your nothingness.
>
> Only so does revelation
> Shine in the time that rejected you.
> Only your nothingness is the experience
> It is entitled to have of you.
>
> (123–4)

In a letter of 17 July 1934, Scholem repeats the key point made in this poem, that Kafka presents 'the world of revelation, but of revelation seen of course from that perspective in which it is returned to its own nothingness' (126). A genuine understanding of Kafka's work must begin, Scholem argues, with this idea of the 'nothingness of revelation'. In response to Benjamin's request that he elaborate on this interpretation, Scholem writes in a letter of 20 September 1934:

> You ask me what I understand by the 'nothingness of revelation'? I understand by it a state in which revelation appears to be without meaning, in which it still asserts itself, in which it has *validity* but no *significance* [*in dem*

sie gilt, aber nicht bedeutet]. A state in which the wealth of meaning is lost and what is in the process of appearing (for revelation is such a process) still does not disappear, even though it is reduced to the zero point of its own content, so to speak. This is obviously a borderline case in the religious sense, and whether it can really come to pass is a very dubious point. (142; Scholem's emphasis)

In an essay on the Kabbalah published in *Judaica* 3, Scholem returns to this image of the borderline (*Grenze*). He clarifies that, in his view, Kafka's work is unsurpassed in its bringing to expression 'the border between religion and nihilism', and that it is the secularized presentation of a Kabbalistic sense of the world (Scholem 1970: 271; my translation).

In a letter to Benjamin dated 26 August 1936, Scholem mentions in relation to his interpretation of Kafka a long essay he has published in the Schocken *Almanach* on 'mystical nihilism' (Scholem 1989: 184). In the eighth of his Hilda Stich Strook Lectures of 1938 (published in 1941 under the title *Major Trends in Jewish Mysticism*), Scholem addresses this question of 'mystical nihilism' in eighteenth-century Sabbatianism. As Scholem explains, this movement believed in the 'holiness of sin', its slogan being that 'We must *all* descend into the realm of evil in order to vanquish it from within. [...] Evil must be fought with evil' (Scholem 1941: 311; Scholem's emphasis). Scholem sees this movement as proposing what he terms a 'nihilist' interpretation of the Torah: 'The Torah, as the radical Sabbatians were fond of putting it, is the seed-corn of Salvation, and just as the seed-corn must rot in the earth in order to sprout and bear fruit, the Torah must be subverted in order to appear in its true Messianic glory' (314). It is a 'mystical nihilism' akin to that of radical Sabbatianism which Scholem finds in Kafka. As his comments to Benjamin indicate, however, he sees Kafka not simply as nihilist – there *is* revelation – but as on its border: the revelation means nothing (*bedeutet nichts*), an expression that echoes Huelsenbeck's on Dada in 1916.

Benjamin's response to Scholem's interpretation of Kafka is to be found in a letter of 20 July 1934, in which he explains that in his own 1934 essay on Kafka, which both repeats and develops arguments made in his 1931 review-essay occasioned by Max Brod's publication of Kafka's posthumous writings, *Beim Bau der chinesischen Mauer*, he has 'endeavoured to show how Kafka sought – on the nether side of that "nothingness", in its inside lining, so to speak – to feel his way toward redemption. This implies that any kind of victory over that nothingness, as understood by the theological exegetes around Brod, would have been an abomination for him' (Benjamin in Scholem 1989: 129). The nothingness present in Kafka's work cannot be denied, then, but neither does it make of that work a simple expression of nihilism, as Lukács will later claim. Rather, the nothingness to which Benjamin refers is required by Kafka for there to be 'the small, nonsensical hope, as well as the creatures for whom this hope is intended and yet who on the other hand are also the creatures in which this absurdity is mirrored' (135).

That Benjamin is nonetheless preoccupied with nihilism in the field of the literary is clear from his comment (in a letter to Scholem dated 2 July 1937) that he intends to write an essay on 'the peculiar figure of medical nihilism in literature: Benn, Céline, Jung' (197). Although this essay was never published, the extensive material that Benjamin collected in the course of the later 1930s in Paris for what has come to be known as *The Arcades Project* (*Das Passagenwerk*) does contain, in convolute K, a note on 'anthropological nihilism' that gives both Louis-Ferdinand Céline and Gottfried Benn as examples. This note includes a reference to the following passage in convolute N regarding 'a specifically clinical nihilism' that was 'first disclosed in explosive fashion by Expressionism' and that is also to be found in the works of Benn and his 'camp follower' Céline (Benjamin 1999a: 472). Basing his argument on remarks made in Carl Gustav Jung's essay 'The Spiritual Problem of Modern Man' (1932), where Expressionism is presented as having taken a 'subjective turn well in advance of the more general change', Benjamin argues that this clinical nihilism 'is born of the shock imparted by the interior of the body to those who treat it' (472). He concludes the fragment with the assertion that 'we should not lose sight of the relations which Lukács has established between Expressionism and Fascism' (472). Hence, by way of Lukács, Benjamin connects fascism with nihilism.

As for Lukács's interpretation of Kafka, it is important to situate this within Lukács's decidedly non-Benjaminian view of aesthetic modernism, to which he is opposed precisely because he takes it to be nihilist. If, in Kafka, this modernist nihilism is at its most extreme, with the transcendence of nothingness, it is nonetheless also present in one form or another, according to Lukács, in the work of all modernist writers, including James Joyce, Robert Musil, William Faulkner, Gottfried Benn, and Samuel Beckett. In the works of these modernists, Lukács finds nihilism above all in what he takes to be their negation of any objective reality, or their 'attenuation of actuality' (Lukács 1963: 25). In each case, the human being becomes an isolated windowless monad trapped within an irredeemably subjective realm: 'Man, for these writers, is by nature solitary, asocial, unable to enter into relationships with other human beings' (19–20). And, according to Lukács, the nihilism of aesthetic modernism thus defined is 'the spontaneous product of the capitalist society in which intellectuals have to live' (91). In short, the nihilism of aesthetic modernism is a direct consequence of the nihilism of capitalist modernity.

According to Marshal Berman, this capitalist nihilism was diagnosed not by Nietzsche but by Marx as early as *The Communist Manifesto* (1848), even if Marx does not actually use the term 'nihilism' at any point. For Marx, the bourgeoisie are:

> the most violently destructive ruling class in history. All the anarchic, measureless, explosive drives that a later generation will baptize by the name of 'nihilism' – drives that Nietzsche and his followers will ascribe to such cosmic traumas as the Death of God – are located by Marx in the seemingly banal everyday working of the market economy. He unveils the modern bourgeois as consummate nihilists on a far vaster scale than modern intellectuals can conceive. (Berman 1983: 100)

From this perspective, the nihilism of the capitalist economy lies in its reduction of all values to exchange value. As Berman puts it: 'This is what modern nihilism is all about. Dostoevsky, Nietzsche and their twentieth-century successors will ascribe this predicament to science, rationalism, the death of God. Marx would say that its basis is far more concrete and mundane: it is built into the banal everyday workings of the bourgeois economic order – an order that equates our human value with our market price' (111). Like Nietzsche, however, Marx finds reason to hope in the very consummation of nihilism. In accordance with a dialectical process, capitalist nihilism leads inexorably to its self-overcoming.

Returning to Lukács, he is far from being alone in his characterization of Kafka's works as nihilist, and he was an important influence on subsequent readings of Kafka. As Peter Heller observed in 1966 of the post-war reception of Kafka: 'in non-Communist Europe the mood and the traditions of Central European *Kulturkritik* and

Kulturpessimismus largely dominated the interpretations of Kafka, and at least in recent decades, the West German critics have been prone to stress the issue of nihilism in Kafka or to suggest an aesthetic perspective which is itself predicated on the claim that nihilism is the "truth" of our age' (Heller 1966: 250). In *The Literature of Nihilism* (1975), Charles I. Glicksberg aims to chart the engagement with nihilism in a literary tradition extending from Turgenev to Beckett, and including Dostoevsky, Leonid Andreyev, Henry de Montherlant, André Malraux, Albert Camus, Jean-Paul Sartre, Eugene Ionesco, and Nikos Kazantzakis. According to Glicksberg – and in this respect he repeats Lukács – Kafka is indisputably 'the most complex and prophetic of the modern nihilists', the 'most gifted and influential exemplar of nihilism in modern literature' (Glicksberg 1975: 14, 124). Unlike Lukács, however, Glicksberg locates Kafka's nihilism in the profound sense of alienation from the society of his time that comes across in his works, and argues that this nihilism is 'a temperamental trait that owes nothing or very little to historical or social conditions' (128). What makes Kafka important as a writer, according to Glicksberg, is not simply that he expresses a 'nihilistic vision', but rather that this vision is *resisted* in his work: 'Though Kafka was an incorrigible skeptic, […] he never gave up the quest for ultimate meaning, impossible as the task proved to be. And it is this steadfast refusal to give in to the nihilism of the absurd that makes him an exemplary figure of our time' (134, 132). Through this combination of a nihilistic vision and a struggle to overcome nihilism, Kafka would resemble Nietzsche, the paradigmatic figure of philosophical modernism, and on this point Glicksberg agrees with Günther Anders, who claims that 'If Nietzsche was not Kafka's teacher he was certainly his prototype in the desperate attempt to overcome nihilism and discover the secret of a new strength of soul, the strength perhaps merely to *be*, without the need of religious meaning' (Anders 1960: 74; Anders's emphasis). The evidence of Kafka's diaries and letters suggests, however, that he was considerably more familiar with the works of Kierkegaard than with those of Nietzsche.

Just as Dada is seen by some commentators as nihilist and by others not simply as non-nihilist but as *resistant* to nihilism or as *anti*-nihilist, so Kafka's works have been far from unanimously interpreted as the expression of a nihilist vision, however that nihilism might be defined. Indeed, Walter Benjamin's attempt to explore the ways in which Kafka articulates a 'small, nonsensical hope' in the face of the seemingly hopeless or the impossible has proven to be a touchstone for an important tradition in Kafka interpretation. Among the most notable readings of this kind are those by the French writer and critic Maurice Blanchot and the German critical theorist Theodor Adorno.

In a series of essays on Kafka, commencing with 'Reading Kafka' (1943) and concluding with 'The Very Last Word' (1968), Blanchot's general argument is that the negativity in Kafka's oeuvre is not nihilist in nature, since it is orientated towards affirmation rather than negation, the latter being the way rather than the goal. In his 1943 essay, for instance, Blanchot declares that 'Kafka's entire work is in search of an affirmation that it wants to gain by negation' (Blanchot 1995b: 7). Ironically, Kafka's works are dark not because there is no hope but precisely because hope survives all failures: 'Kafka's narratives are among the darkest in literature, the most rooted in absolute disaster. And they are also the ones that torture hope the most tragically, not because hope is condemned but because it does not succeed in being condemned' (10). One might even go so far as to say that Blanchot's Kafka is a writer in whose works nihilism fails, Nietzsche's 'death of God' resulting not in a sense of absence, nothingness, or despair, but rather in the spectrality of the divine or what in his 1943 essay Blanchot describes as 'dead transcendence': 'The dead God has found a kind of impressive revenge in this work. For his death does not deprive him of his power, his infinite authority, or his infallibility; dead, he is even more terrible, more invulnerable, in a combat in which there is no longer any possibility of defeating him. It is a dead transcendence we are battling with' (7). The negativity specific to Kafka, then, would be non-dialectical, a negativity in the wilderness that fails to accomplish that clearing of the ground promised by various forms of nihilism. Blanchot characterizes such negativity as radically, irreducibly ambiguous – 'The ambiguity of the negation is linked to the ambiguity of death. God is dead, which may signify this harder truth: death is not possible' (7) – and he sees this ambiguity as the defining characteristic of literature as such.

Anticipating Adorno's take on Kafka, Blanchot identifies Hunter Gracchus as the paradigmatic figure in Kafka's mature work: a personage whose death has somehow miscarried and who inhabits a space between life and death, between being and non-being. In one of the posthumously published 'Gracchus' fragments, Gracchus recounts from the ship of death:

> 'I have lain here ever since the time when, as the Hunter Gracchus living in the Black Forest, I followed a chamois and fell from a precipice. Everything happened in good order. I pursued, I fell, bled to death in a ravine, died, and this ship should have conveyed me to the next world. [...]
> 'I had been glad to live and I was glad to die. Before I stepped aboard, I joyfully flung away my wretched load of ammunition, my knapsack, my hunting rifle that I had always been proud to carry, and I slipped into my winding sheet like a girl into her marriage dress. I lay and waited. Then came the mishap.' (Kafka 1993: 369–70)

As a result of this mishap, Gracchus now inhabits an afterlife which is this side of death: 'I am here, more than that I do not know, further than that I cannot go. My ship has no rudder, and it is driven by the wind that blows in the undermost regions of death' (370–1).

In his essay 'Kafka and Literature' (1949), Blanchot expands on his earlier characterization of Kafka's work as radically ambiguous, declaring that his art – and indeed literature as such – is 'the place of anxiety and complacency, of dissatisfaction and security. It has a name: self-destruction, infinite disintegration. And another name: happiness, eternity' (Blanchot 1995b: 26). In the section on Kafka in *The Space of Literature* (1955), he defines art in relation to what the poet Friedrich Hölderlin, in his elegy 'Bread and Wine' (1801), terms the 'destitute time' (*dürftige Zeit*) of the gods' absence: 'It describes the situation of one who has lost himself, who can no longer say "me", who in the same movement has lost the world, the truth of the world, and belongs to exile, to the *time of distress* when, as Hölderlin says, the gods are no longer and are not yet' (Blanchot 1982: 75). As we have seen in the previous chapter, this conception of modernity also lies at the heart of Heidegger's thoughts on art's relation to nihilism, although Heidegger makes no mention of Kafka. Because this *dürftige Zeit* is 'never without hope' in Kafka, one can assert with confidence, according to Blanchot, that the 'nihilistic perspective' has been 'too hastily attributed to him' (76).

Like Blanchot, although with a more explicit commitment to locating Kafka historically in relation to specific socio-economic and political forces, Adorno also challenges the reading of his oeuvre as nihilist. In his 1953 essay 'Notes on Kafka', for instance, Adorno claims that 'To include him among the pessimists, the existentialists of despair, is as misguided as to make him a prophet of salvation. He honoured Nietzsche's verdict on the words optimism and pessimism' (Adorno 1981: 269). Adorno's Kafka is a writer who anticipates the experience of the Nazi concentration camps – as the consummation of nihilism – in his depiction of an existence located in the no man's land between life and death, an existence captured most explicitly in the Hunter Gracchus fragment:

> In the concentration camps, the boundary between life and death was eradicated. A middleground was created, inhabited by living skeletons and putrefying bodies, victims unable to take their own lives, Satan's laughter at the hope of abolishing death. As in Kafka's twisted epics, what perished there was that which had provided the criterion of experience – life lived out to its end. Gracchus is the consummate refutation of the possibility banished from the world: to die after a long and full life. (260)

Adorno's Kafka is, then, like Blanchot's, a writer of failed negation. This very failure becomes the sole form that can legitimately be taken by hope in the dark times of modernity.

For both Blanchot and Adorno, Kafka is to be placed alongside Nietzsche rather than Hegel in his writing of the negative, as a writer whose modernism lies precisely in his critique of modernity and its core values: reason, individual freedom, progress. The key difference between Kafka and Nietzsche would lie, however, in the nature of their responses to the perceived nihilism of modernity rather than in the fact that one is a writer of literary texts and the other is a philosopher. This view becomes explicit in Peter Heller's *Dialectics and Nihilism* (1966):

> The attempt to renounce all aspirations to meaning and value is of course relevant to major aspects of Kafka as it is relevant to Nietzsche. But Nietzsche's response to this encounter with nothingness was a frenzied activism, the demand for ever new struggles of creative destruction and destructive creativity imposing ever new structures of autonomous value and meaning upon the self-devouring cycles of life. Kafka's response was a quiet despair – the correlative of his persistence in the search for meaning and value, and to his constant sense of inevitable failure in this search. However, neither Nietzsche nor Kafka recommended complacent accommodation to nihilism, a withdrawal into the shelter of autonomous nonsense, or the pretense that the mere sense of futility may be rephrased as positive gospel. (Heller 1966: 256–7)

Heller's Kafka is close to both Blanchot's and Adorno's in his being a writer of the neither-nor, the liminal space between the living and the dead, being and non-being, presence and absence, meaning and meaninglessness. As Heller puts it, Kafka's oeuvre dramatizes 'on all levels the attempt and the failure to arrive at either adequate meaning or conclusive meaninglessness' (276); it 'dramatizes conclusive inconclusiveness in hovering between meaning and meaninglessness, a despairing movement on the brink of total negation, a state of suspension between life and death' (288). This liminal space is produced textually not only at the level of content – most explicitly in the Hunter Gracchus fragment – but also rhetorically and stylistically, through a writing that is also always an unwriting, a writing that repeatedly takes back what it gives, tempts the reader with, and also resists, an allegorical interpretation. As Adorno puts it: 'Each sentence says "interpret me", and none will permit it' (Adorno 1981: 246).

Ironically, both for those intent on making of Kafka a nihilist and for those who defend his work against this charge, the text now generally known as the *Zürau Aphorisms* (but which was originally published posthumously by Max Brod under the title *Reflections on Sin, Suffering, Hope and the True Way*) has been a key resource. Written in 1917–18, while Kafka was staying in Zürau, a village in north-west Bohemia, and at a time when the pulmonary tuberculosis that would be the cause of his death in 1924 had declared itself, these aphorisms certainly emphasize the role of the negative. Indeed, Glicksberg even goes so far as to claim that they 'embody Kafka's nihilistic outlook' (Glicksberg 1975: 130). One finds, for instance, the claim that 'To perform the negative is what is still required of us, the positive is already ours', and that 'There is a goal, but no way: what we call a way is hesitation' (Kafka 1973: 88). To this might be added the diary entry from 1920, included in the posthumously published selection entitled 'He', in which Kafka states that the writer's aim should be to reveal life 'as a dream, as a hovering, as a nothingness' (Kafka 1990: 855; quoted in Heller 1966: 233). There are, however, numerous aphorisms that do not fall neatly within any conception of nihilism. For instance: 'The Indestructible is one; it is each individual human being and at the same time it is common to all, hence the unparalleled strength of the bonds that unite mankind' (Kafka 1973: 95).

Turning to Kafka's works more generally, it is certainly possible to argue for a nihilism in his representation of the individual's relation to others, and above all the relation to forms of paternal authority. As has often been noted by commentators, Kafka's oeuvre exhibits a preoccupation with mysterious figures of authority who judge – and who often condemn to death – for reasons that remain obscure and seemingly illegitimate, as though the Law were inconsistent, self-flouting, or even mad. In the early story *The Judgement* (written in 1912; published in 1916), this authority figure is a literal father, who passes suddenly from a state of geriatric weakness to one of absolute power, and who condemns his son, Georg Bendemann, to death by drowning, the son then obediently carrying out this sentence on himself. In *The Metamorphosis* (also written in 1912; published in 1915), it is again the father who, switching suddenly from a state of apparent weakness to one of terrorizing strength, literally attacks the son; and again, the son dies. In *The Trial* (written in 1914; published posthumously in 1925), this authority figure is no longer the literal father but rather the Law (*das Gesetz*). As is suggested in the parable 'Before the Law' (included towards the end of the novel and also published separately in 1919 in the collection *A Country Doctor*), this Law remains beyond the reach of the one who is judged by it. On this occasion, the condemned man, Josef K., refuses to carry out the sentence on himself, but he dies all the same – and, as it seems to him, 'Like a dog!' (Kafka 1992b: 251). In Kafka's last, unfinished novel, *The Castle* (which he began in 1922 and which was published posthumously in 1926), the authority figure is even less concrete, although the overall power structure remains the same as in the earlier works. In each of these narratives, Kafka's protagonist is unable to assume power, unable either to master or to escape a world in which, in most cases, he is adjudged to be unworthy of life. Each work charts a failure to achieve mastery of a particular situation; each presents authority as located in a mysterious, unpredictable, and inaccessible elsewhere; each concerns itself with weakness, guilt, and failure. Almost all human relationships are structured hierarchically, and, as Anna Katharina Schaffner has observed (see Schaffner 2010), these relationships are generally sado-masochistic in nature. Far from locating themselves outside these hierarchies, Kafka's protagonists actively participate in them.

If one turns to the form of Kafka's prose, one finds a recurrent rhetorical mode whereby what is posited is then qualified or even retracted. One of the most striking instances of this rhetorical procedure occurs in the exchange recorded by Max Brod in which Kafka refers explicitly to nihilism. The conversation took place on

28 February 1920, and is recorded in Brod's 1937 biography of Kafka. It took the following form:

> [Kafka:] 'We are nihilistic thoughts that came into God's head.' I [Brod] quoted in support the doctrine of the Gnostics concerning the Demiurge, the evil creator of the world, the doctrine of the world as a sin of God's. 'No,' said Kafka, 'I believe we are not such a radical relapse of God's, only one of his bad moods. He had a bad day.' 'So there would be hope outside our world?' He smiled, 'Plenty of hope – for God – no end of hope – only not for us.' (Brod 1947: 61)

Here, Kafka first affirms that there is hope – indeed, 'Plenty of hope' – but then immediately qualifies this affirmation with a negation: 'only not for us'. The *Zürau Aphorisms* contain numerous examples of just such an affirming-negating gesture. For instance:

> The more horses you put to the job, the faster it goes – that is to say, not the tearing of the block out of its base, which is impossible, but the tearing apart of the straps and as a result the gay empty ride. (Kafka 1973: 91)

> Truth is indivisible, hence it cannot recognise itself; whoever wants to recognise it must be a lie. (96)

> You can hold yourself back from the sufferings of the world, this is something you are free to do and it accords with your nature, but perhaps this very holding back is the one suffering that you could avoid. (101)

Arguably the most perfect instance of such a process of affirmation-negation is the following aphorism from the same collection:

> The crows maintain that a single crow could destroy the heavens. There is no doubt of that, but it proves nothing against the heavens, for heaven simply means: the impossibility of crows. (89)

Here, the affirmative moment is itself negative in nature – the possibility of destroying heaven – and the negative moment negates this negativity. It is on this account that the aphorism has been interpreted as stating the impossibility of nihilism. As Peter Heller notes, for instance, in *Kafka oder die unzerstörbare Hoffnung* (1955) Robert

Rochefort proposes an allegorical reading, with Kafka's crows as 'radical nihilists' who seek to 'reveal the world as nothing' (Heller 1966: 232–3). According to this reading, the aphorism states that such nihilism can do nothing to negate the divine, since the latter is the negation of nihilism. Another, arguably more persuasive, way in which to read these aphorisms, however, is to concentrate on the manner in which the rhetoric articulates radical paradoxes and aporias, a kind of hermeneutic paralysis the thematic equivalent of which would be the figure of Hunter Gracchus.

Among the many examples of negation working at even the most micrological level in Kafka's texts is his use of the prefix 'un-'. A particularly striking instance of this is to be found in the famous opening sentence of *The Metamorphosis* (1915). The standard English translation of this sentence reads: 'As Gregor Samsa awoke one morning from uneasy dreams he found himself transformed in his bed into a gigantic insect' (Kafka 1993: 75). The original German reads: 'Als Gregor Samsa eines Morgens aus *unruhigen* Träumen erwachte, fand er sich in seinem Bett zu einem *ungeheureren Ungeziefer* verwandelt' (Kafka 1994: 115; emphasis added). What disappears in the English translation is precisely the recurrence of the 'un-' prefix. While this negating prefix survives in the translation of *unruhig* as 'uneasy', it is lost in the translation of *ungeheuer* as 'gigantic' (the German word also suggesting the monstrous and unfamiliar) and of *Ungeziefer* as 'insect' (the German word in fact being considerably less precise than that, denoting rather a vermin or pest, and constituting a major stumbling block for any would-be translator of the story). *Unruhig – ungeheuer – Ungeziefer*: this sequence captures at the level of the letter the negativity at work in the story, a negativity that will result in Gregor's mistreatment and then complete rejection by his family, and ultimately in his death.

As Anna Katharina Schaffner has observed (see Schaffner 2010), another important instance of such negativity working through the triple deployment of the 'un-' prefix is to be found in the late, unfinished story 'The Burrow' (written in 1923; first published in 1931), about a subterranean creature that feels itself threatened by an undefined enemy, and that seeks to protect itself through the construction of an elaborate burrow. Reflecting on his dream of those 'structural devices' that would enable him 'to slip in and out at will', the creature asks himself: 'what does it amount to?' (*was soll es?*), and answers his own question: 'It is the mark of a restless frame of mind, of inner uncertainty, of disreputable desires, bad qualities' (Kafka 1973: 205). The original German here reads: 'Es deutet auf *unruhigen* Sinn, auf *unsichere* Selbsteinschätzung, auf *unsaubere* Gelüste, schlechte Eigenschaften [...]' (Kafka 1992a: 482; emphasis added). The adjective *unruhig* ('uncalm' or 'unquiet') is, as we have seen, the first 'un-' word to appear in *The Metamorphosis* and a frequent word in 'The Burrow'); while *unsauber* ('unclean') connects back to *Ungeziefer*, the latter originally referring to an animal considered unfit for sacrifice on account of its being unclean (see Corngold 1996: 87). Here, too, a sequence of 'un-' words produces a negativity at the level of the letter which complements that at the level of content. Indeed, it is significant that this sequence of negative prefixes should occur in relation to a reflection on 'structural devices' (*Baumöglichkeiten*) that would enable the creature to 'slip in and out unobserved', since the text itself is constructed in a manner akin to that of the burrow (*Bau*), labyrinthine and at once refuge and trap.

One finds, then, that the negative is operational in Kafka both thematically and rhetorically. On this basis, one might argue that his work is nihilistic. At the same time, however, it is possible to point to various ways in which this negativity appears to be countered within the texts. Anticipating the reading of Gilles Deleuze and Félix Guattari (1973), Max Brod emphasizes the humour in Kafka, arguing that 'Even the most gruesome episodes in Kafka's writings [...] stand in a curious twilight of humour, an investigator's interest and tender irony. This humour, which is an essential ingredient of Kafka's writing (and of his manner of living), points through the meshes of reality to the divine existence beyond' (Brod 1947: 42). And between these two positions – Kafka as nihilist, Kafka as non-nihilist – there is a third position, namely that Kafka's work concerns itself with a failure of the negative that is not in itself positive. Peter Heller takes this third position when he argues that Kafka's oeuvre is the consummation of an anti-Faustian tradition, a tradition that rejects the principle of striving (*Streben*) or, more precisely, seeks to reveal that all striving is in vain: 'Kafka's work is an attempt to contradict, to terminate, to reverse the tradition. He reveals only the impossibility of transcendence within the temporal sphere, and thus the futility of striving' (Heller 1966: 303). This position is essentially the one taken by Roger Griffin when he characterizes Kafka's work as the prime instance of 'epiphanic modernism', a modernism of withdrawal or resignation. According to Nietzsche, however, this kind of withdrawal would itself be a form of nihilism – namely 'passive nihilism' of the kind that he finds in Schopenhauer's conception of the denial of the will to live (*Verneinung des Willens zum Leben*) and also in Christianity.

What this suggests, of course, is that all depends on one's definition of nihilism, and that those who either charge Kafka with nihilism (as does Lukács) or who defend him against that charge (as do Blanchot and Adorno) do so for reasons that relate directly to their own conceptions of the literary and its relation to nihilism, rather than to anything essentially nihilistic in Kafka's work. What is shared by each of the above commentators on Kafka, however, is the conviction that the literary has a key – indeed, *the* key – role to play in the struggle against nihilism. For Lukács, it is necessary to critique the aesthetic modernism that finds its consummation in Kafka precisely because, in Lukács's view, another form of the literary – nineteenth-century realism of the kind epitomized by Balzac in the novels and stories making up his *Human Comedy* and carried into the twentieth century by Thomas Mann – is the true counterforce to nihilism, offering readers an objective vision of the world in which the human being is a social and historical entity. For Blanchot and Adorno, on the other hand, the negativity to be found in Kafka makes of his work the true counterforce to nihilism, since it constitutes a resistance to the central values of modernity as established in the French Revolution: not only liberty, equality, and fraternity, but also reason and progress. For Adorno, these values have proven themselves, in accordance with the 'dialectic of enlightenment', to have led to the horrors of totalitarianism, and are countered by an experience of radical ambiguity and the aporetic of the kind articulated by Kafka.

Among the writers of the 'high' modernist period between the two world wars, Kafka's oeuvre stands out as paradigmatic with respect to the relation between aesthetic modernism and nihilism precisely on account of its constituting the site of such warring interpretations. On the one hand, his oeuvre is taken to be the most extreme expression of nihilism. On the other hand, it is seen as the privileged form of resistance to the nihilism of modernity. Similarly antithetical positions have been taken on major writers who come after Kafka and who fall not only within a broadly conceived aesthetic modernism but also within a tradition that finds its point of origin in Kafka. It is to these writers that we may now turn.

Post-war literatures of the unword: Beckett, Blanchot, Celan

In the post-Second World War period, the work of a number of major writers has attracted comparison with Kafka's, and in each case the question of nihilism has proven to be central in their critical reception. Similarities between Kafka's works and those of Samuel Beckett (1906–89) were noted in the early 1950s, when Beckett was acquiring an international reputation following the publication of the three novels *Molloy* (1951), *Malone Dies* (1951), and *The Unnamable* (1953), and the staging of the play *Waiting for Godot* (France, 1953; Great Britain, 1955; the United States, 1956). In *The Meaning of Contemporary Realism* (1957), Georg Lukács placed Beckett alongside Kafka, arguing that in the former one finds 'a fully standardized nihilistic modernism' (Lukács 1963: 53). Similarly, reviewing *Molloy* in April 1951, Maurice Nadeau argued that the novel is 'full of a somewhat Kafkaesque brand of humor', that the human being in Beckett's work is 'thrown into a meaningless world', and that 'Beckett settles us in the world of the Nothing where some nothings which are men move about for nothing' (Nadeau in Graver and Federman 1979: 52–3). A year later, reviewing *Malone Dies*, Nadeau claimed that, in this novel, metaphysics 'is very concrete and explosive, even merry. It proclaims the nothingness of life, the nothingness of man; it moves in an absolute nihilism' (78). Other early reviewers of Beckett's first post-Second World War works reached a similar conclusion. Bernard Pingaud declared in a September 1951 review of *Molloy* that Beckett 'is undoubtedly obsessed by the idea of death and nothingness' (Pingaud in Graver and Federman 1979: 68), and, across the Channel, Philip Toynbee, reviewing the English translation of *Molloy* in December 1955, claimed that Beckett was 'the end-product of a fictional tradition which has flowed from Kafka through Sartre, Camus, and [Jean] Genet, and of a tradition of nihilistic writing which goes back to [Alfred] Jarry, to Lautréamont, to Sade. What he has done is to carry his despair and disgust to ultimate limits of expression – indeed beyond them' (Toynbee in Graver and Federman 1979: 74). This view of Beckett recurs in later commentaries on his work, with Glicksberg, for instance, arguing that Beckett is 'the literary nihilist *par excellence*, though he does not call himself by that name' (Glicksberg 1975: 235).

On the one hand, then, Beckett's work has been seen as the consummation of a nihilistic modernism established by Kafka. On the other hand, however, as with Kafka, so with Beckett, there is a strong critical counter-tradition to this view. According to Adorno, for instance, Beckett's oeuvre is like Kafka's precisely in its retention of a 'haven of hope' in the liminal space between being and nothingness, the space inhabited by Kafka's Hunter Gracchus and also by Beckett's vagabond protagonists. As Adorno puts it in the section on 'Nihilism' in *Negative Dialectics* (1966):

> To Beckett, as to the Gnostics, the created world is radically evil, and its negation is the chance of another world that is not yet. As long as the world is at it is, all pictures of reconciliation, peace, and quiet resemble the picture of death. The slightest difference between nothingness and coming to rest would be the haven of hope, the no man's land between the border posts of being and nothingness. (Adorno 1973: 381)

Adorno's defence of Beckett as non-nihilist does not take the form of an appeal to the positive, however. For, as Adorno sees it, any such appeal would itself be a nihilist gesture. Paradoxically, the nihilists are those 'who oppose nihilism with their more and more faded positivities, the ones who are thus conspiring with the extant malice, and eventually with the destructive principle itself. Thought honors itself by defending what is damned as nihilism' (381). This defence of works charged with nihilism takes the form of a reading of Beckett that insists upon a 'slightest difference' within the negative. What Adorno means by this is clarified by his comments on Beckett in his 1965 lecture series on 'Metaphysics' and by the draft fragments towards an unwritten essay on *The Unnamable* included in Adorno's own copy of the 1959 German translation of the novel. In the 1965 lectures, Adorno argues that Beckett's oeuvre 'revolves around the question what nothingness actually contains; the question, one might say, of a topography of the void. This work is really an attempt so to conceive nothingness that it is, at the same time, not *merely* nothingness, but to do so within complete negativity' (Adorno 2000: 135–6; Adorno's emphasis). This notion of a nothingness that is at odds with itself in Beckett is developed in the following notes in Adorno's copy of the German translation of *The Unnamable*:

> the positive categories, such as hope, are the absolutely negative ones in B[eckett]. Hope is directed at nothingness.

Is nothingness the same as nothing? [*Ist das Nichts gleich nichts?*] That is the question around which everything in B[eckett] revolves. Absolutely everything is thrown away, because there is only hope where nothing is kept back. The fullness of nothingness. This is the reason for the insistence on the zero point. (186 n. 15)

Adorno's defence of Beckett against the charge of nihilism is, then, one that insists upon a difference within the negative. A very different approach is taken by the French philosopher Alain Badiou, who declares that he is 'entirely opposed to the widely held view that Beckett moved towards a nihilistic destitution, towards a radical opacity of significations' (Badiou 2003: 55), and that he 'will *never* be a nihilist' (15). Rather than nihilism, what Badiou finds in Beckett's work is 'a powerful love for human obstinacy, for tireless desire, for humanity reduced to its stubbornness and malice' (75); in short, a commitment to going 'on'. Like Badiou, Simon Critchley dismisses as the 'stalest of all the stale philosophical clichés' the claim that Beckett's oeuvre 'celebrates the meaninglessness of existence and is therefore nihilistic' (Critchley 1997: 176). According to Critchley, what Beckett offers is 'an approach to meaninglessness as an achievement of the ordinary without the rose-tinted glasses of redemption, an acknowledgement of the finiteness of the finite and the limitedness of the human condition' (179). The position of one of the major French philosophers of the post-war period, Jacques Derrida, is distinct from those outlined above in that he sees Beckett's works as being *both* 'nihilist' *and* 'not nihilist', the nihilism being located in the content and the non-nihilism in the 'composition, the rhetoric, the construction and rhythm of his works' (Derrida 1992: 61).

Those who seek to defend Beckett against the charge of nihilism would appear to be in line with Beckett's own stated position. Whereas Kafka claims that 'We are nihilistic thoughts in the mind of God', Beckett is reported by the literary critic Gottfried Büttner as having observed on 10 September 1967: 'I simply cannot understand why some people call me a nihilist. There is no basis for that' (Beckett quoted in Büttner 1984: 122). To this may be added Beckett's comment to the French writer Charles Juliet during a conversation on 11 November 1977 that 'Negation is no more possible than affirmation. It is absurd to say that something is absurd. That's still a value judgement. It is impossible to protest, and equally impossible to assent' (Juliet 1995: 165). It would be a mistake, however, simply to accept this view of the matter, not least because Beckett also offered as a gloss on his work the following summary of the thought of the pre-Socratic philosopher Gorgias, who is described as a nihilist in Archibald Alexander's *Short History of Philosophy* (1907), the work from which Beckett derived this summary:

> Three celebrated propositions –
>
> 1. Nothing exists.
> 2. If it did, it could not be known.
> 3. If it could be known, it could not be communicated.
>
> (quoted in Feldman 2006: 76)

That the 'nothing' occupies a central place in Beckett's work is clear from his comments in a letter of 9 July 1937 to Axel Kaun on a proposed 'literature of the unword'. Such a literature would be one in which language effaces itself in order to disclose that which lies behind it. Beckett's conception of a self-undoing language is articulated as follows:

> It is indeed getting more and more difficult, even pointless, for me to write in formal English. And more and more my language appears to me like a veil which one has to tear apart in order to get to those things (or the nothingness) lying behind it. […] It is to be hoped the time will come, thank God, in some circles it already has, when language is best used where it is most efficiently abused. Since we cannot dismiss it all at once, at least we do not want to leave anything undone that may contribute to its disrepute. To drill one hole after another into it until that which lurks behind, be it something or nothing, starts seeping through – I cannot imagine a higher goal for today's writer. (Beckett 2009a: 518)

Two days later, in a letter to Mary Manning Howe, Beckett claims that he is starting a 'Logoclasts' League', which will produce a 'ruptured writing, so that the void may protrude, like a hernia' (521).

In his first published novel, *Murphy* (1938), which he completed the year before the letter to Kaun, Beckett writes of an experience of 'the Nothing, than which in the guffaw of the Abderite naught is more real' (Beckett 1938: 246). Here, Beckett is alluding to the pre-Socratic philosopher Democritus, a summary of whose philosophy he had read in Alexander's *Short History of Philosophy*. This summary includes the following passage:

> Aristotle, in his account of the early philosophers, says, 'Leucippus and Democritus assume as elements the "full" and the "void". The former

they term being and the latter non-being. Hence they assert that non-being exists as well as being.' And, 'according to Plutarch, Democritus himself is reported as saying, 'there is naught more real than nothing'. (Alexander 1922: 38–9)

In *Murphy*, the experience of this nothing is presented as an all-too-fleeting pleasure. Beckett's next novel, *Watt* (completed in 1945; published in 1953), also contains an experience of the nothing. On this occasion, however, the one who experiences the nothing (Watt) finds it profoundly troubling, since it resists both rational explanation and linguistic expression. During Watt's time in the house of Mr Knott, two men – 'the Galls father and son' – come to tune the piano:

> What distressed Watt in this incident of the Galls father and son, and in subsequent similar incidents, was not so much that he did not know what had happened, for he did not care what had happened, as that nothing had happened, that a thing that was nothing had happened, with the utmost formal distinctness, and that it continued to happen, in his mind, he supposed, though he did not know exactly what that meant [...]. (Beckett 2009c: 62–3)

Watt finds this happening of 'nothing' unbearable, since it defies both reason and expression: 'For the only way one can speak of nothing is to speak of it as though it were something' (64).

The thought of the nothing – either as that which is desired or as that which is feared – recurs throughout Beckett's subsequent works. Just as Hemingway has a character rewrite the paternoster in a nihilist manner, so Beckett has Moran in the novel *Molloy* (1951) recite 'the pretty quietist Pater, Our Father who art no more in heaven than on earth or in hell, I neither want nor desire that thy name be hallowed' (Beckett 2009b: 175). In *Malone Dies* (1951), Beckett returns to the Democritean idea of the nothing found in *Murphy* when Malone reflects: 'I know those little phrases that seem so innocuous and, once you let them in, pollute the whole of speech. *Nothing is more real than nothing*' (Beckett 1959: 193; Beckett's emphasis). *Waiting for Godot* (1952) is punctuated by the refrain 'Nothing to be done' (*Rien à faire*). Clov, in the play *Endgame* (1957), declares: 'Better than nothing! Is it possible?' (Beckett 1990: 121). The radio play *Embers* (1959) ends with the protagonist, Henry, consulting his diary and saying: 'Saturday ... nothing. Sunday ... Sunday ...

nothing all day. [*Pause.*] Nothing all day. [*Pause.*] All day all night nothing' (Beckett 1990: 264). The late prose text *Ill Seen Ill Said* (1981) ends: 'Grace to breathe that void. Know happiness' (Beckett 1982: 59). And one of Beckett's last works, *Worstward Ho* (1983), articulates what Beckett describes homophonically as a 'gnawing to be naught' (Beckett 1983: 46).

Those commentators who have sought to defend Beckett against the charge of nihilism have generally placed the emphasis on his works' recurrent insistence upon the need to 'go on' despite all failures to establish a meaning in what the character Molloy describes as 'senseless, speechless, issueless misery' (Beckett 2009b: 10). Indeed, not only do they find a kind of affirmation that would prevent Beckett's oeuvre from being the expression of either ontological or ethical nihilism, but they often see it, for all its apparent negativity, as countering nihilism. What such a reading tends to obscure, however, is the manner in which nihilism of various kinds – ontological, ethical, aletheiological – haunts that oeuvre. And it is just such a haunting that can be found in the work of the other major writers within the tradition extending back to Kafka.

The fiction of the French writer and critic Maurice Blanchot (1907–2003) bears more than a passing resemblance to Beckett's. Indeed, Blanchot even records that Beckett 'was willing to recognize himself' in Blanchot's 1962 work *Awaiting Oblivion* (Blanchot 1995a: 299). Like Beckett's, too, Blanchot's work has been charged with nihilism. In his highly influential book *The Rhetoric of Fiction* (1961), the American literary critic Wayne C. Booth devotes a short section to nihilism in literature. His analysis includes the comment: 'I am told by Marcel Gutwirth [...] that the novels of Maurice Blanchot come remarkably close to a thoroughgoing nihilism' (Booth 1961: 298 n. 22). The philosopher Emmanuel Levinas – who, unlike Booth, was familiar both with Blanchot's fiction and with his critical writings – states in an interview published in 1971 that Blanchot's oeuvre 'can be interpreted in two directions at the same time'. The first of these directions is to see it as:

> the announcement of a loss of meaning, a scattering of discourse, as if one were at the extreme pinnacle of nihilism – as if nothingness itself could no longer be thought peacefully, and had become equivocal to the ear listening to it. Meaning, bound to language, in becoming literature, in which it should be fulfilled and exalted, brings us back to meaningless

repetition – more devoid of meaning than the wandering structure or piecemeal elements that might make it up. (Levinas 1996: 154)

According to Levinas, however, the second direction leads to a sense of transcendence in the work, and indeed 'more transcendence than any world-behind-the-worlds ever gave a glimpse of' (154–5).

We have seen that Blanchot's reading of Kafka seeks to counter the interpretation of the latter's works as nihilist. In his critical writings more generally, Blanchot repeatedly insists that his own conception of literature is not nihilist. In the introduction to his first collection of essays, *Faux Pas* (1943), he identifies what he takes to be the 'naïve calculation' of the nihilist, and argues that the 'hope of the nihilist – to write a work, but a destructive work' is 'foreign' to the genuine writer, since the latter 'obeys anguish, and anguish commands him to lose himself, without this loss being compensated by any positive value' (Blanchot 2001: 6–7). Blanchot's writer is in the 'ludicrous condition of having nothing to write, of having no means with which to write it, and of being constrained by the utter necessity of always writing it' (3). The 'nothing' (*rien*) is, then, that which has to be written; and, precisely because this 'nothing' cannot be written, writing finds itself caught up in an interminable task. Throughout his work, Blanchot figures this interminable task as the impossibility of death or endless dying. He repeatedly insists, however, that this experience of death's impossibility is not nihilist. In one of his most important essays, 'Literature and the Right to Death' (1948), for instance, Blanchot returns to the central argument in the introduction to *Faux Pas* – that a certain 'nothing' speaks in literature – and clarifies precisely this point:

> a nothing demands to speak, nothing speaks, nothing finds its being in speech, and the being of speech is nothing. This formulation explains why literature's ideal has been the following: to say nothing, to speak in order to say nothing. That is not the musing of a high-class kind of nihilism. (Blanchot 1995b: 324)

This distinction between literature and nihilism remains in place throughout Blanchot's later critical writings.

The notion of death's impossibility, or the impossibility of writing the nothing, or of accomplishing a Hegelian negation that would result in the sublation (*Aufhebung*) of that which is negated, is a recurrent theme in Blanchot's fiction, and it also shapes the style of his writing. The narrator of his story *The Madness of the Day* (1949) recounts the experience of near-execution: 'the madness of the world broke out. I was made to stand against the wall like many others. Why? For no reason. The guns did not go off. I said to myself, God, what are you doing?' (Blanchot 1981: 6). A similar experience is recounted, on this occasion in the third person and within a concrete historical setting, in his last work, *The Instant of My Death* (1994). Here, the scene is set during the German occupation of France, and those responsible for the execution are German soldiers. Blanchot writes:

> I know – do I know it – that the one at whom the Germans were already aiming, awaiting but the final order, experienced then a feeling of extraordinary lightness, a sort of beatitude (nothing happy, however) – sovereign elation? The encounter of death with death?
>
> In his place, I will not try to analyze. He was perhaps suddenly invincible. Dead – immortal. Perhaps ecstasy. Rather the feeling of compassion for suffering humanity, the happiness of not being immortal or eternal. (Blanchot 2000: 5)

Again, however, this anticipation of not being immortal is unfulfilled; the execution does not take place, and the brief narrative ends: 'What does it matter. All that remains is the feeling of lightness that is death itself or, to put it more precisely, the instant of my death henceforth always in abeyance' (11).

Figures who inhabit a liminal space between life and death, between being and non-being, recur throughout Blanchot's fiction, and these recall Kafka's Hunter Gracchus. Indeed, Gracchus's statement 'I had been glad to live and I was glad to die' (Kafka 1993: 369) is echoed in the opening paragraph of Blanchot's *The Madness of the Day*: 'I am alive, and this life gives me the greatest pleasure. And what about death? When I die (perhaps any minute now), I will feel immense pleasure' (Blanchot 1981: 5). Such figures may also be related to what, in his aphorisms, Kafka names 'the indestructible' (*das Unzerstörbare*), that which resists all possible negation and thus renders nihilism unrealizable, as Blanchot argues in his essay 'Crossing the Line' (1958), the title of which alludes to texts by Ernst Jünger (1950) and Martin Heidegger (1955) on the possibility of overcoming nihilism:

> Until now we thought nihilism was tied to nothingness. How ill-considered that was: nihilism is tied to being. Nihilism is the

impossibility of being done with it and of finding a way out even in that end that is nothingness. It says the impotence of nothingness, the false brilliance of its victories; it tells us that when we think nothingness we are still thinking being. [...] Nihilism tells us its final and rather grim truth: it tells of the impossibility of nihilism. (Blanchot 1993: 149)

The claim made in the above text is striking because, in redefining nihilism as the 'impotence of nothingness', Blanchot suggests that the experience of the impossibility of death that recurs throughout his fiction is in fact the experience of nihilism. However, this view has to be set alongside Blanchot's ongoing attempt to distinguish his own thinking of the negative from the 'naïve calculation' of the nihilist. Taking these two positions together leads neither to an identification of Blanchot's fiction as nihilist nor to an identification of it as non- or anti-nihilist. Rather, nihilism may be said to haunt his fiction in a manner closely akin to its haunting of Beckett's oeuvre. Blanchot's own repeated attempts to present literature as the counterforce to nihilism not only reveal his place within a tradition that can be traced back to Nietzsche but also serve to mask that haunting which makes nihilism, in Nietzsche's words, the 'uncanniest of all guests'.

We have seen that Kafka is a spectral presence in the works of both Beckett and Blanchot, and that a haunting of those works by the thought of nihilism may be related to that presence by way of Kafka's claim in his *Zürau Aphorisms* that 'To perform the negative is what is still required of us'. In the work of the Romanian-born German-language poet Paul Celan (1920–70), one finds a similar conjunction. Here, however, that conjunction is explicitly connected with the Holocaust. For Adorno – and he is far from being alone in this – the Holocaust is the consummation of the nihilism of modernity, and its aftermath a time when, as Adorno puts it in his 1965 lectures on metaphysics, it is impossible to insist 'on the presence of a positive meaning or purpose in being. [...] To assert that existence or being has a positive meaning constituted within itself and orientated towards the divine principle (if one can put it like that), would be, like all the principles of truth, beauty and goodness which philosophers have concocted, a pure mockery in face of the victims and the infinitude of their torment' (Adorno 2000: 101–2).

The question then becomes whether Celan's poetry can speak of the Holocaust in a manner that is not itself nihilist. Celan's own conception of poetry 'after Auschwitz' certainly appears to come very close to being nihilist. In his most extensive poetological text, *The Meridian* (1962), Celan defines poetry (*Dichtung*) as 'this declaring endless of mere mortality and the in-vain' (*diese Unendlichsprechung von lauter Sterblichkeit und Umsonst*) (Celan 1992: iii. 200; my translation). And his friend the German-Jewish academic Peter Szondi claims that 'death, the memory of the dead, lies at the origin of Celan's entire poetic oeuvre' (Szondi 2003: 32). In *The No One's Rose*, the collection of poems that Celan published a year after *The Meridian*, the emphasis upon the negative is particularly strong, nowhere more so than in the poem 'Psalm', from which the title of the entire collection is taken. The poem reads as follows:

> No one moulds us again out of earth and clay,
> no one conjures our dust.
> No one.
>
> Praised be your name, No One.
> For your sake
> we shall flower.
> Towards
> you.
>
> A nothing
> we were, are, shall
> remain, flowering:
> the nothing-, the
> No one's rose.
>
> With
> our pistil soul-bright,
> with our stamen heaven-ravaged,
> our corolla red
> with the crimson word which we sang
> over, O over
> the thorn.
>
> (Celan 2002: 153)

To this poem, it might seem only reasonable to apply Georg Lukács's interpretation of Kafka's work in general: that there is a 'transcendence of Nothingness' and that 'If there is a God here, it can only be the God of religious atheism: *atheos absconditus*.' Moreover, the 'only purpose of transcendence – the intangible *nichtendes Nichts*' would be 'to reveal the facies hippocratica of the world' (Lukács 1963: 44–5). Here, however, the God who has become 'No one' (*Niemand*) is the God who did not save the European Jews from the Holocaust, and the poem may be read as a reproach delivered by those who are now 'A nothing' (*Ein Nichts*). On the one hand, then, the poem would appear to be performatively nihilist. As John Felstiner observes, 'the opening words annul not only comfort but Creation itself' (Felstiner 1995: 168). On the other hand, the poem speaks of that which remains: the 'no one's rose' (*Niemandsrose*) will continue to flower; the prayer will continue to be uttered, albeit to a God become 'No one'. It is upon this remaining (*restance*) that the philosopher Jacques Derrida insists when he argues that 'To address no one is not exactly not to address any one. To speak to no one, *risking*, each time, singularly, that there might be no one to bless, no one who can bless – is this not the only chance for blessing? for an act of faith? What would a blessing be that was sure of itself? A judgment, a certitude, a dogma' (Derrida 2005c: 42). For another commentator, Shira Wolosky, the 'Nothing' in this poem – and also in the poem 'Mandorla' in the same collection, in which the answer to the question 'what stands in the almond?' is 'The nothing' – 'signals at once nihilism and messianism' (Wolosky 1995: 256). As we shall see in the next chapter, nihilism and messianism have been thought together in a tradition extending back to Walter Benjamin and Gershom Scholem in the modern period. As for Celan's own view on the relation of his work to nihilism: on the one hand, as noted above, he presents poetry as a 'declaring endless of mere mortality and the in-vain'; on the other hand, as he writes in a letter of 10 August 1962 to Erich Einhorn concerning the poem 'Stretto', in the collection *Speech-Grille*, that work 'evokes the devastation caused by the atom bomb. At a central place stands, as fragment, this sentence by Democritus: "There exists nothing except atoms and empty space; everything else is opinion." I don't need to underline that the poem was written because of that opinion – for the sake of the *human*, thus against all emptiness and atomizing' (Celan 2005: 181–2; Celan's emphasis).

A form of nihilism appears, then, to haunt Celan's work just as it haunts both Kafka's and Beckett's. That both Kafka and Beckett are relevant here is indicated not least by the fact that Celan considered writing a doctoral thesis on Kafka and quotes one of his *Zürau Aphorisms* – 'Never again psychology!' (Kafka 1991: 96) – in one of his later poems, 'Frankfurt, September', in the collection *Threadsuns* (1968), and that he is reported as having said shortly before he committed suicide in April 1970 that Beckett was probably the only person with whom he might have had a mutual understanding (see Wurm 1995: 250). To suggest that the works of these three writers are haunted in a particular way by nihilism is not to suggest, however, that their works *are* nihilist, or even that they *express* nihilism at one level and counter it at another (that is, in their style), as Derrida claims of Beckett. Rather, nihilism – in what Nietzsche terms its 'uncanniness' – is located on a border such that it is neither inside nor outside their work, and for this reason cannot simply be dismissed or exorcized from that work. It is this haunting that is missed by those commentators who assert either that Kafka and those who follow in his wake are nihilist and also by those who assert that each of these writers produces an art that resists nihilism. While the reception of their works has certainly been shaped by Nietzsche's conception of modernity as nihilist and by his claim that art is the 'only superior counterforce' to this nihilism, it has not taken account of Nietzsche's own insistence upon the uncanniness of nihilism, an uncanniness that renders hazardous any attempt to deploy the concept of nihilism in order either to critique or to champion aesthetic modernism. With this in mind, we can now turn to the postmodern engagement with the concept of nihilism, and consider whether a clear distinction can be made between modernist and postmodernist conceptions of nihilism, and whether the latter does take adequate account of the uncanniness of nihilism.

Part III
POSTMODERNISM AND NIHILISM

5

OUR ONLY CHANCE?

We have seen that, in both philosophical and aesthetic modernism, nihilism is repeatedly taken to be that which has to be resisted or overcome. For these forms of modernism, nihilism is the essence of a modernity against which modernism turns in either a programmatic or an epiphanic manner. If there is a distinctly *postmodern* attitude to nihilism, then its specificity lies in a revalorization of nihilism. Complicating any simple distinction between modernism and postmodernism in their respective attitudes to nihilism is the fact that there are instances of just such a revalorization in modernism itself. Within the realm of philosophico-political modernism, nihilism is championed by Cloots at the time of the French Revolution, and by Russian nihilists such as Pisarev and Nechaev in the 1860s. Within the realm of aesthetic modernism, a revalorization of nihilism occurs in some strains of Dada and in Elliot Paul's celebration of what he terms the 'new nihilism' of a writer such as Drieu La Rochelle. Such revalorizations are, however, decidedly rare within both philosophical and aesthetic modernism, where, as we have seen, the emphasis falls much more often upon the threat posed by nihilism and the manner in which this threat is to be countered.

Just as, for the most part, both philosophical and aesthetic modernism define themselves in antagonistic relation to a variously conceived nihilism, so there are strains of what is generally taken to be postmodern thought in which nihilism continues to be seen either as that which has to be resisted or as that against which a given form of thought defines itself. This attitude towards nihilism is to be found in the work of one of the key theorists of the postmodern,

Jean-François Lyotard, who in *The Postmodern Condition* (1979) argues that it is in Kant's theory of the sublime rather than in Nietzsche's conception of nihilism that the essence of modernity is captured:

> Modernity, in whatever age it appears, cannot exist without a shattering of belief and without discovery of the 'lack of reality' of reality, together with the invention of other realities.
>
> What does this 'lack of reality' signify if one tries to free it from a narrowly historicized interpretation? The phrase is of course akin to what Nietzsche calls nihilism. But I see a much earlier modulation of Nietzschean perspectivism in the Kantian theme of the sublime. I think in particular that it is in the aesthetic of the sublime that modern art (including literature) finds its impetus and the logic of avant-gardes finds its axioms. (Lyotard 1984: 77)

While Lyotard does not go as far as to set the postmodern in antagonistic relation to nihilism, many other thinkers who associate themselves either with postmodernism or with poststructuralism do just this. In 'Politics and the Limits of Modernity' (1988), for instance, the post-Marxist political theorist Ernesto Laclau characterizes postmodern politics as beyond all foundationalism, and, in so doing, insists that such a politics is not to be mistaken for nihilism:

> Abandonment of the myth of foundations does not lead to nihilism, just as uncertainty as to how an enemy will attack does not lead to passivity. It leads, rather, to a proliferation of discursive interventions and arguments that are necessary, because there is no extradiscursive reality that discourse might simply reflect. Inasmuch as argument and discourse constitute the social, their open-ended character becomes the source of a greater activism and a more radical libertarianism. Humankind, having always bowed to external forces – God, Nature, the necessary laws of History – can now, at the threshold of postmodernity, consider itself for the first time the creator and constructor of its own history. The dissolution of the myth of foundations – and the concomitant dissolution of the category 'subject' – further radicalizes the emancipatory possibilities offered by the Enlightenment and marxism. (Laclau 1988: 79–80)

Such attempts to think the postmodern in contradistinction to nihilism have prompted repeated attacks by thinkers who take postmodernism and/or poststructuralism to be unambiguously nihilist in essence. These attacks have come from various quarters, philosophical, political, aesthetic, and religious, and tend to present postmodernism as a relativism insisting on a multiplicity of views, none of which possesses greater validity than any other, or as a denial of all Enlightenment values, including not only universality but also reason and the idea of progress. Attacks of this kind often come from defenders of Christianity as the religion of the West. Andrew Harrigan, for instance, claims that:

> The essence of post-modernism is nihilism – the denial of any meaning or purpose in existence – or, more exactly, the triumph of nihilism in societies of the Western world. It is a phenomenon identical with atheism as it denies the existence of any permanent ethical order. It marks a turning away from the moral teachings that have come down to us from Moses and are brought to their highest level in Christianity. Nihilism is worse than paganism because the pagans knew no better. It is worse than barbarism, for the barbarians of antiquity at least based their lives on tribal rules. (Harrigan 1998: 24)

Within the more limited sphere of poststructuralism in literary studies, the charge of nihilism is generally made along lines similar to those adopted by René Wellek, who in 'The New Nihilism in Literary Studies' claims that 'No self, no author, no coherent work, no relation to reality, no correct interpretation, no distinction between art and nonart, fictional and expository writing, no value judgment, and finally no truth, but only nothingness – these are negations that destroy literary studies' (Wellek 1990: 80).

In *The Aesthetics of Disappearance* (1980), the French philosopher Paul Virilio proposes a rather different conception of postmodernity as nihilist. Contesting Heidegger's conception of nihilism in terms of technology, Virilio argues that 'The nihilism of technique destroys the world less surely than the nihilism of speed destroys the world's truth' (Virilio 1991: 69). According to Virilio, it is speed in the postmodern world that is annihilating, collapsing any distinction between here and there, now and then: 'Reconciliation of nothingness and reality, the annihilation of time and space by high speeds substitutes the vastness of emptiness for that of the exoticism of travel, which was obvious for people like Heine who saw in this very annihilation the supreme goal of technique' (109).

As for the attitude adopted by those postmodernist or poststructuralist thinkers against whom the charge of nihilism is directed, Karen L. Carr claims in *The Banalization of Nihilism* (1992) that they exhibit 'a kind of indifferent acceptance, as though, properly understood, nihilism is not a significant concern and should not be made out as such' (Carr 1992: 105). In fact, nothing could be further from the truth. For instance, the majority of those philosophers and literary theorists who associate themselves with deconstruction are anything but indifferent to the question of nihilism. Indeed, the attempt to defend themselves against the charge of nihilism is a recurrent gesture on their part.

The 'great question': deconstruction and nihilism

In 'The Critic as Host' (1979), the literary theorist J. Hillis Miller defines deconstruction as 'neither nihilism nor metaphysics but simply interpretation as such, the untangling of the inherence of metaphysics in nihilism and of nihilism in metaphysics by way of the close reading of texts' (Hillis Miller 1979: 230). As the 'untangling' of this inherence, deconstruction 'does not provide an escape from nihilism, nor from metaphysics, nor from their uncanny inherence in one another. There is no escape' (231). Instead, deconstruction 'makes the inherence oscillate in such a way that one enters a strange borderland, a frontier region which seems to give the widest glimpse into the other land ("beyond metaphysics"), though this land may not by any means be entered and does not in fact exist for Western man' (231).

The insistence that deconstruction is not itself nihilist is even more pronounced in the work of that movement's founder, Jacques Derrida (1930–2004). Time and again, Derrida states in the most unambiguous fashion that he considers it a complete misunderstanding of deconstruction to see it as a form of nihilism. In a 1984 interview with Richard Kearney, for instance, he declares: 'I totally refuse the label of nihilism which has been ascribed to me and my American colleagues. Deconstruction is not an enclosure in nothingness, but an openness towards the other' (Derrida 1984: 124). In *Memoires: For Paul de Man* (1986), on the literary theorist whose work is in many respects most closely aligned with Derrida's, he writes: 'I especially wish to denounce the sinister ineptitude of an accusation – that of "nihilism" – which so many major professors, following the example of minor journalists, have often made against Paul de Man and his friends' (Derrida 1989: 21). In 'Paul de Man's War' (1988) – an essay prompted by the disclosure of de Man's having published numerous articles in two Belgian newspapers, *Le Soir* and *Het Vlaamsche Land*, between 1940 and 1942, during the German occupation of Belgium and at a time when these publications were controlled by the occupying forces – Derrida even goes so far as to claim that the commentators who charge de Man with nihilism are using a term that was deployed by the Nazis and those who collaborated with them:

> the accusation of 'nihilism', often directed helter-skelter against de Man or against deconstruction in general, not only testifies both to the non-reading of texts and to a massive lack of sensitivity to the great question – still open and still redoubtable – of nihilism and of metaphysics. This accusation bespeaks either political amnesia or a lack of political culture. Those who toss around the word nihilism so gravely and so lightly should, however, be aware of what they're doing: under the occupation, the 'propagators' of dangerous ideas were often denounced by accusing them of 'nihilism', sometimes in violently antisemitic tracts, and always in the name of a new order, moral and right-thinking […]. (261)

When, during an interview in 2002, Derrida was asked what he considered to be 'the most widely held misconception' about him and his work, he responded:

> That I'm a skeptical nihilist who doesn't believe in anything, who thinks nothing has meaning, a text has no meaning. That's stupid and utterly wrong, and only the people who haven't read me say this. This misreading of my work began 35 years ago and it's very difficult to destroy. […] Anyone who reads my work with attention understands that I insist on affirmation and faith, and that I'm full of respect for the texts I read. (Derrida 2005a: 121–2)

Such remarks make it very clear that, far from being indifferent to nihilism, Derrida is at pains to defend his own work against a charge that he places in the historical context of Nazi rhetoric during the Second World War.

If Derrida belongs squarely within philosophical modernism in his taking of nihilism to be that against which his own thought is most forcefully opposed, his conception of nihilism is in one key respect closer to Adorno's than it is to Nietzsche's or to Heidegger's. Just as Adorno sees nihilism as that which negates difference in the interests of what in *Negative Dialectics* (1966) he terms 'absolute integration', so Derrida takes nihilism to be that which negates the other in its otherness. This is made clear in the above-cited declaration that 'Deconstruction is not an enclosure in nothingness, but an openness towards the other.' Such a conception of nihilism is also operative in Constantin Boundas's theorization of what he terms a 'minoritarian deconstruction'. As Boundas puts it: 'From the minoritarian deconstructive perspective, nihilism is the assimilation of difference, that is, the appropriation of the other, instead of the becoming-other' (Boundas in Darby et al. 1989: 82). Whereas nihilism, as Derrida understands it, negates the other, annihilates difference in the name of identity, deconstruction aims to disclose and preserve differences; it is 'a positive response to an alterity which necessarily calls, summons or motivates it' (Derrida 1984: 118).

In an interview with Maurizio Ferraris published under the title 'I Have a Taste for the Secret' in 2001, Derrida describes that negation of alterity which he elsewhere terms 'nihilism' as 'bad violence'. This form of violence is, he argues, 'impoverishing, repetitive, mechanical'; it 'does not open the future, does not leave room for the other'; it is a violence that 'homogenizes and effaces singularity' (Derrida 2001: 92). For Derrida, as for Adorno before him, nihilism as 'bad violence' finds its most extreme incarnation in Nazism. In *Force of Law* (1994), Derrida takes up Walter Benjamin's concept of 'mythological violence' – as defined in the latter's 1921 essay 'Critique of Violence' – and applies it to Nazism as '*the final achievement of the logic of mythological violence*' (Derrida 2002: 295; Derrida's emphasis). This mythological violence negates the other rather than opening up a space for it. If there is a categorical imperative guiding the counter-violence of deconstruction, then it is that there should be an other 'rather than nothing', as Derrida puts it in the interview with Ferraris. The ethics of deconstruction lie precisely in this anti-nihilist affirmation of alterity:

> I must have had occasion to say, for example, that it's better that there be a future, and that I move in the direction of deconstruction because it is what comes, and it's better that there be a future, rather than nothing. For something to come there has to be a future, and thus *if there is* a categorical imperative, it consists in doing everything for the future to remain open. I am strongly tempted to say this, but then – in the name of what would the future be worth more than the past? More than repetition? Why would the event be preferable to the non-event? Here I might find something that resembles an ethical dimension, because the future is the opening in which the other happens, and it is the value of the other

or of alterity that, ultimately, would be the justification. Ultimately, that is my way of interpreting the messianic. The other may come, or he may not. I don't want to programme him, but rather to leave a place for him to come if he comes. It is the ethic of hospitality. (Derrida 2001: 82–3; Derrida's emphasis)

As a practice of reading, the deconstructive affirmation of the other takes the form of what Derrida terms a 'countersigning' of the text being read. This countersigning does not seek to master the text, is not an attempt to exhaust it through the disclosure of its meaning or meanings. Rather, as Derrida puts it in an interview with Evelyne Grossman in 2003, 'the duty of the reader-interpreter is to write while letting the other speak, or so as to let the other speak' (Derrida 2005c: 166–7). Deconstruction might be described, then, as hospitable reading.

If Derrida follows the major philosophical modernists in his redetermination of nihilism and in his conception of philosophy's task as the resistance of that nihilism, he is no less closely aligned with them in the privilege that he accords to literature as that which can resist nihilism by opening a space for alterity. That Derrida takes a particular kind of literary practice to be resistant to nihilism is made clear in one of his earlier books, *Of Grammatology* (1967), where he presents modernist literary practice as 'dislocating' the 'founding categories of language and the grammar of the *épistémè*', with Stéphane Mallarmé's poetics being 'the first break in the most entrenched Western tradition' (Derrida 1976: 92). As Timothy Clark has observed, Mallarmé is not simply one writer among others for Derrida, for it is with Mallarmé that literature in Derrida's 'limited, specialised sense' first emerges (Clark 1992: 17–18). The conception of literature that Derrida traces back to Mallarmé is one that makes it a 'privileged guiding thread' in the 'modern period'. The reason for this privilege is what happens to language in the literary work:

> What literature 'does' with language holds a revealing power which is certainly not unique, which it can share up to a point with law, for example with judicial language, but which in a given historical situation (precisely our own, and this is one more reason for feeling concerned, provoked, summoned by 'the question of literature') teaches us more, and even the 'essential', about writing in general, about the philosophical or scientific (for example linguistic) limits of the interpretation of writing. (Derrida 1992: 71–2)

For Derrida, as for Maurice Blanchot before him, Mallarmé's work is distinctive in that the text's relation to meaning and to reference is suspended (see Derrida 1992: 47). Crucially, however, this suspension of meaning does not result in meaninglessness – that is, in nihilism. Put slightly differently, the literary as Derrida defines it turns back upon itself, and in so doing departs or differs from itself. Kafka's 'Before the Law' (1919) is, for Derrida, a literary event of this kind, since Kafka's text 'points obliquely to literature, speaking of itself as a literary effect – and thereby exceeding the literature of which it speaks' (Derrida 1992: 215). Or, as Derrida puts it in *Demeure: Fiction and Testimony* (1998), in which he proposes a reading of Maurice Blanchot's *The Instant of My Death* (1994): 'There is no essence or substance of literature: literature is not. It does not exist. It does not remain at home, *abidingly* [à demeure] in the identity of a nature or even of a historical being identical with itself' (Derrida 2000: 28; Derrida's emphasis). It is, then, not simply that literature 'is not', but rather that the literary text remarks upon its not being what it is – this is, precisely, literature's 'revealing power'. Literature, as Derrida conceives it, not only differs from itself, but also *discloses* that differing, reveals the alterity at the heart of the same. It is precisely on account of this 'revealing power' that literature is, according to Derrida, to be accorded a privilege in the struggle against the 'bad violence' of nihilism defined as the annihilation of alterity.

While the idea that the literary resists nihilism through its openness to the other underlies all of Derrida's texts on a range of literary figures within the field of European aesthetic modernism – including Mallarmé, Artaud, Joyce, Kafka, Blanchot, and Genet – it is nowhere more evident than in his writings on Paul Celan, where the poet's language is set in direct opposition to the language of nihilism. In an interview published under the title 'Language is Never Owned' in 2000, for instance, in which he discusses his various readings of Celan, Derrida claims that the specificity of the poetic act lies in its 'resurrection' of language beyond the various deaths to which it is subjected, including the death imposed upon it by Nazism, which Derrida describes as 'a crime against the German language' (Derrida 2005c: 106). If, for Heidegger, *Dichtung* names Being and thereby resists nihilism as the forgetting of Being (*Seinsvergessenheit*), the literary as Derrida conceives it is that event which opens onto the other, saves language from the annihilating principle of identity. This 'other' can take many forms – although, significantly, it would seem that it cannot take the form of the 'nothing'. This limitation becomes apparent in an interview with Derek Attridge, in which Derrida defines the literary event as the 'nothing-ing of nothing' (Derrida 1992: 47).

The privilege that Derrida accords to the literary for its 'nothing-ing of nothing' – that is, for its countering of nihilism – aligns him with the key figures of philosophical modernism, including Nietzsche, Heidegger, and Adorno, and militates against the definition of deconstruction as a philosophical postmodernism that can be clearly distinguished from philosophical modernism. Like Adorno before him, however, Derrida substitutes the 'other' for 'Being' as that which must be saved if the nihilism of modernity is to be resisted. This substitution marks Derrida's departure from Heidegger's philosophical modernism, but, framed as it is within a thinking of nihilism as the 'great question', it does not mark his departure from philosophical modernism as such. That this is the case is evident above all in the privilege that Derrida accords to the literary in the struggle against nihilism.

'Messianic nihilism': the return to Walter Benjamin

One of the key differences between the late work of Paul de Man, the American literary theorist most closely associated with deconstruction, and that of Derrida is the former's readiness to gesture towards a revalorization of nihilism. In an early review-essay entitled 'The Literature of Nihilism' (1966), de Man defends romanticism against the charge of nihilism directed at it by Erich Heller on account of what the latter takes to be its neo-Hellenism. De Man argues that most of the romantics moved beyond what Heller terms the 'nostalgic envy' of ancient Greece, and that in the case of Friedrich Hölderlin there was never any such 'mood of regret' (de Man 1989: 169). If romanticism is nihilist, then paradoxically the 'literature of nihilism is not necessarily nihilistic' (164). This idea of a non-nihilistic nihilism returns in one of de Man's last works, his essay on Walter Benjamin's 'The Task of the Translator' (1923). Here, in the discussion following the delivery of this paper at Cornell University in March 1983, de Man refers to another essay by Benjamin, the 'Theologico-Political Fragment' (written, according to Benjamin's editors, either in 1920–1 or in 1937–8). Benjamin's essay ends with the claim that the 'task of world politics' is to achieve the complete 'passing away' (*Vergängnis*) of nature, including 'those stages of man that are nature'. According to Benjamin, the method of such a politics 'must be called nihilism' (Benjamin 2002: 306). De Man takes up this idea and connects it with his own claim that literature offers a 'negative knowledge' of the ideological, naturalizing function of language:

> One could say, with all kinds of precautions, and in the right company, and with all kinds of reservations, that – and I think that's a very small company – that Benjamin's concept of history is nihilistic. Which would have to be understood as a very positive statement about it. In the same way that in Nietzsche nihilism is a necessary stage, and is accounted for in those terms. Understand by nihilism a certain kind of critical awareness which will not allow you to make certain affirmative statements when those affirmative statements go against the way things are. (de Man 1986: 103–4)

The revalorization of nihilism in Benjamin's fragment is also invoked by another major post-war thinker, the Italian philosopher Giorgio Agamben. Taking as his point of departure Walter Benjamin's concept of 'bare/mere life' (*bloßes Leben*) in the essay 'Critique of Violence' (1921), Agamben argues that the essence of Western politics is biopolitics, and that this biopolitics is founded on an originary exclusion of 'bare life' from the political. However, this very exclusion of 'bare life' renders it that which may be destroyed without this destruction being subject to the law. According to Agamben, Nazi Germany was the 'first radically biopolitical state' in this sense, with the extermination camps being the place where 'bare life' was annihilated (Agamben 1998: 143). Agamben sees Western biopolitics leading towards a 'biopolitical catastrophe' even more extreme than the one brought about by Nazism (188). Such a biopolitics is, for Agamben, nothing other than a form of political nihilism, and this nihilism cuts across all party-political distinctions. Invoking Gershom Scholem's remark to Walter Benjamin that the relation to the law in Kafka's *The Trial* is the 'Nothing of Revelation' (*Nichts der Offenbarung*), Agamben proceeds to argue that:

> All societies and all cultures today (it does not matter whether they are democratic or totalitarian, conservative or progressive) have entered into a legitimation crisis in which law (we mean by this term the entire text of tradition in its regulative form, whether the Jewish Torah or the

Islamic Shariah, Christian dogma or the profane *nomos*) is in force as the pure 'Nothing of Revelation'. But this is precisely the structure of the sovereign relation, and the nihilism in which we are living is, from this perspective, nothing other than the coming to light of this relation as such. (51)

In the face of such political nihilism, which he terms 'imperfect nihilism', Agamben proposes not an anti-nihilism but rather a counter-nihilism, or a 'perfect nihilism'. Like de Man, Agamben turns for the theorization of such a counter-nihilism to Benjamin's conception of 'messianic nihilism' as articulated in his epistolary exchanges with Gershom Scholem in the 1930s. In 'The Messiah and the Sovereign: The Problem of the Law in Walter Benjamin' (1992), Agamben elaborates on this notion of a messianic nihilism as follows:

> If we accept the equivalence between messianism and nihilism of which both Benjamin and Scholem were firmly convinced, albeit in different ways, then we will have to distinguish two forms of messianism or nihilism: a first form (which we may call imperfect nihilism) that nullifies the law but maintains the Nothing in a perpetual and infinitely deferred state of validity, and a second form, a perfect nihilism that does not even let validity survive beyond its meaning but instead, as Benjamin writes of Kafka, 'succeeds in finding redemption in the overturning of the Nothing'. (Agamben 1999: 171)

According to Agamben, a politics of imperfect nihilism cannot master the originary philosophico-political 'fracture' in the concept of life as both *zoē* and *bios*. Nazism, as the first radically biopolitical programme, 'darkly and futilely sought to liberate the political scene of the West from this intolerable shadow [the Jews as "bare life"] in order to produce the German *Volk* as the people that finally overcame the originary biopolitical fracture' (Agamben 1998: 179). Not only did this programme fail, but Nazism actually reproduced the fracture with the entire German people as an instance of 'bare life'; that is, 'a sacred life consecrated to death' (180). Hence, the consummation of Nazism was its own self-annihilation, along with that of the German people.

As I have sought to demonstrate elsewhere (see Weller 2008: 137–62), Agamben's distinction between imperfect and perfect nihilism, with the latter serving as a form of revalorized counter-nihilism, is beset with a number of problems, not the least of which is that this very distinction between two kinds of nihilism constitutes one more 'fracture' of the kind that he is aiming to critique. That said, Agamben's return to Benjamin for a revalorization of nihilism locates him within a more general postmodern tendency which finds its most explicit articulations in the work of the philosophers Jean Baudrillard and Gianni Vattimo, to whom we may now turn.

The nihilism of the transparent society: Baudrillard

Whereas Derridean deconstruction is aligned with other forms of philosophical modernism in its identification of nihilism as that against which it struggles in the name of the 'other', the work of the French philosopher Jean Baudrillard (1929–2007) might seem to be more properly postmodern in its take on nihilism, if the postmodern is understood as involving either a revalorization of, or an indifference to, the concept of nihilism. In his essay 'On Nihilism', in *Simulacra and Simulation* (1981), Baudrillard proposes a tripartite history of nihilism, in which its first great manifestation is late eighteenth- and early nineteenth-century romanticism. That movement was nihilist, according to Baudrillard, on account of its 'destruction of the order of appearances' (Baudrillard 1994: 159). In an argument that echoes Max Weber's theory of *Entzauberung*, Baudrillard sees this destruction as a radical 'disenchantment of the world', a shift towards 'the violence of interpretation and of history' (160). The second great manifestation of nihilism occurs with the avant-garde movements of the early decades of the twentieth century, and in particular Dada and Surrealism. Whereas romanticism destroys the order of appearances, the avant-garde destroys 'the order of meaning'. Whereas romantic nihilism is aesthetic, avant-garde nihilism is political, historical, and metaphysical, even if, as discussed in Chapter 3 above, it continues to privilege the aesthetic.

The third manifestation of nihilism – which might be described as the postmodern manifestation, although Baudrillard does not use that term – is distinct from the romantic and avant-garde manifestations in that it cannot be grasped in Nietzschean terms; that is, as 'decadence', the 'death of God', or even the 'violence of interpretation'. Rather, this third manifestation has to be understood as 'simulated transparency', in which everything becomes a simulacrum (*simulacre*), a copy without original, a surface that hides no depth. According to Baudrillard, this third form of nihilism is 'more radical, more crucial than in its prior and historical forms, because this transparency, this irresolution is indissolubly that of the system, and that of all the theory that still pretends to analyze it' (159). In other words, this form of nihilism is more radical because it is all-pervasive: there is no outside from which it might be critiqued. Furthermore, not only does it destroy history, the social, the individual, and the real, but this third manifestation of nihilism also destroys the concept of nihilism as such. The theory of nihilism is no longer appropriate to describe simulated transparency, since it remains 'a desperate but determined theory, an imaginary of the end, a weltanschauung of catastrophe' (161). In its most extreme form, then, nihilism annihilates even itself:

> in disappearance, in the desertlike, aleatory, and indifferent form, there is no longer even pathos, the pathetic of nihilism – that mystical energy that is still the force of nihilism, of radicality, mythic denial, dramatic anticipation. It is no longer even disenchantment, with the seductive and nostalgic, itself enchanted, tonality of disenchantment. It is simply disappearance. (162)

Baudrillard seeks to distinguish his conception of nihilism from the various nihilisms within philosophical modernism, above all Nietzsche's and Heidegger's. He does so not only by defining nihilism as the destruction of the real, but also by revalorizing nihilism. If there is no position outside nihilism from which it might be critiqued, and if one is not simply to submit to the prevailing nihilism of the transparent society, then, Baudrillard argues, one has to turn nihilism back against itself. Hence his claim: 'I am a nihilist' (160). While resorting to Nietzsche's concept of 'active nihilism' to characterize the kind of nihilism that he himself embraces, Baudrillard nonetheless concludes that even such active nihilism plays into the hands of the prevailing nihilism of the transparent society, becoming 'the involuntary accomplice of the whole system, not politically, but in the accelerated form of indifference that it contributes to imposing' (163). The indifference to which Baudrillard is referring here is a 'generalized process of indifferentiation' that is the essence of the contemporary mediatic age (161). As a mediatic rather than an aesthetic, political, historical, or metaphysical phenomenon, postmodern nihilism renders real suffering indifferent. An example of this indifferentiation is the manner in which the violence of war becomes a spectacle on the television screen, and one that does not traumatize the viewer. On the one hand, then, Baudrillard champions active nihilism in a Nietzschean manner as that which promises the overcoming of nihilism. On the other hand, this active nihilism belongs to the nihilist 'system', and can achieve nothing. It is within this circuit that Baudrillard's thinking of nihilism remains caught.

Nihilism as 'aesthetic consciousness': Vattimo

For a philosophical approach that explicitly sets out to revalorize nihilism rather than to demonize it (as does Derrida), but that does not also see the revalorization of nihilism as itself futile (as does Baudrillard), one has to turn to the work of the Italian philosopher Gianni Vattimo. In a series of works including *The Adventure of Difference* (1980), *The End of Modernity* (1985), and the essays collected in *Nihilism and Emancipation* (2003), Vattimo undertakes a revalorization of nihilism as what he terms 'our (only) chance' (Vattimo 1988: 23). Through readings of Nietzsche and Heidegger, Vattimo redetermines nihilism as the negation of all metaphysical ground or foundation. This, he argues, is what Nietzsche means when he announces the 'death of God'. Although that foundation may have been named in various ways – including not only God, but also truth, reality, reason, fact, and objectivity – the master term for the foundation that is negated by nihilism is Being, and it is here that Heidegger's determination of nihilism as the forgetting, abandonment, or default of Being comes into play.

According to Vattimo, philosophical postmodernism begins not with post-Second World War thought but with Nietzsche, since it is Nietzsche who inaugurates that epoch in the history of Being wherein Being itself is submitted to nihilism as a process of weakening (*indebolimento*). Vattimo sees this weakening of Being continuing in Heidegger's thought. In neither Nietzsche nor Heidegger, however, does this nihilistic weakening result in an annihilation of Being. Such an annihilation would be a reactive or bad nihilism, from which Vattimo wishes to distance the active nihilism that he champions. Nihilism, he argues, 'remains ensnared in metaphysics as long as it conceives of itself, even only implicitly, as the *discovery* that there, where we thought *there was* Being, *there is in reality* nothing' (Vattimo 2004: 146; Vattimo's emphasis).

At the heart of Vattimo's revalorization of nihilism, then, lies Nietzsche's crucial distinction between two basic kinds of nihilism: active and reactive. It is active nihilism that is 'our (only) chance', because it undoes the ground upon which any kind of fundamentalism – philosophical, political, or religious – could be based. All such fundamentalisms are, for Vattimo, forms of unethical violence. Active nihilism is postmodern in that it is a hermeneutic violence orientated not towards the disclosure of Truth, but rather towards the disclosure of *truths,* and in this it is ethical violence. The self-undoing ground for such an ethical violence is Nietzsche's assertion in his 1886–7 notebook that there are no facts, only interpretations (see Nietzsche 2003: 139). However, active nihilism thus conceived has to be thought not as an overcoming (*Überwindung*) of the kind theorized by Nietzsche, but rather as what Heidegger terms a *Verwindung*. In *The End of Modernity*, Vattimo seeks to clarify the difference between the two key concepts of *Überwindung* and *Verwindung*. While *Verwindung* 'indicates something analogous' to *Überwindung*, it is distinct from the latter 'both because it has none of the characteristics of a dialectical *Aufhebung* and because it contains no sense of a "leaving-behind" of a past that no longer has anything to say to us' (Vattimo 1988: 164).

Whereas *Überwindung* is modernist, *Verwindung* is a postmodernist form of active nihilism. Whereas active nihilism affirms that there are no facts, only interpretations, reactive nihilism refuses to accept that 'neither *objective* meanings and values nor *given* structures of Being exist' (Vattimo 2006: 135; Vattimo's emphasis). Whereas active nihilism commits itself to the 'creative task' of producing 'new values and new structures of meaning, new interpretations' (135), reactive nihilism is non-creative and clings to old values that are rooted in those given structures of Being. As a *Verwindung* of Being, as distinct from its overcoming, however, active nihilism cannot simply free itself from reactive nihilism: active nihilism is 'always passive and reactive too' (140). As a *Verwindung*, active nihilism is the interminable undoing of reactive nihilism, an 'indefinite process of reduction, diminution, weakening' (Vattimo 1998: 93). As we shall see, this insistence on interminability is central not only to Vattimo's conception of nihilism, but also to various forms of aesthetic postmodernism.

On the one hand, the epoch of active nihilism thus conceived is simply one in a series of historical epochs, since 'there is no *Grund* or ultimate truth; there are only historically destined or historically dispatched overtures from a *Selbst* or Same, which gives itself to us only in and through these overtures' (Vattimo 1988: 175). On the other hand, the epoch of Being in which active nihilism is the overture is distinct from all others, since it marks the final (if endless) phase in the history of Being:

> Nihilism, if it should (and can) not be understood as the discovery that instead of Being there is nothing, can only think itself as the history (endless – without conclusion in a state in which in place of Being *there is* nothing) in which Being, asymptotically, consumes itself, dissolves, grows weak. (Vattimo 2004: 146; Vattimo's emphasis)

Where Vattimo's theorization of active nihilism rejoins Enlightenment thought, and thus the very modernity from which it aims to take its distance, is in his conviction that this series of epochs in the history of Being is not arbitrary or chaotic, but progressive. On the one hand, Vattimo wishes to break with the metaphysical philosophy of history that 'dominates the Enlightenment and grounds its faith in the widening cone of light cast by reason' (153). On the other hand, he offers a vision of history in which the postmodern *Verwindung* of modernity through active nihilism carries the West towards an ever more just, ever more ethical, ever less violent condition. Vattimo thus contrasts his own conception of active nihilism with the nihilism he finds in the thought of Gilles Deleuze, which he characterizes as 'a nihilism that has no hope of constructing anything historically and no prospect of becoming a "state"' (Vattimo 2006: 201).

If Vattimo remains aligned with philosophical modernism in his determination of his own 'weak ontology' as working against a non-constructive nihilism, he is also aligned with it in the privilege that he accords to the aesthetic in the resistance of reactive nihilism. Vattimo's attitude to Derridean deconstruction might lead one to conclude that in fact his view of the aesthetic is precisely what separates him from Nietzsche's claim that art is the 'only superior counterforce' to nihilism. In 'Philosophy and the Decline of the West' (1998), for instance, Vattimo criticizes Derrida for producing a discourse that is 'poetic' rather than 'rational' (Vattimo 2004: 24). Vattimo even goes so far as to describe Derridean deconstruction as being, like the neopragmatism of Richard Rorty, a 'pure esthetic game' (29).

However, Vattimo's distinction between his own brand of active nihilism and deconstruction is questionable for a number of reasons.

Just as Derrida appeals to an ethic of hospitality and the value of the other when seeking to justify deconstruction, so Vattimo appeals to an 'ethics of the other, or the others' (64). Just as Derrida distinguishes between the ethical violence of deconstruction and the 'bad violence' of nihilism, so Vattimo distinguishes between the justifiable violence of active nihilism and the unacceptable violence of reactive nihilism. Just as Derrida claims in *Force of Law* that deconstruction is itself justice, so Vattimo claims in *Nihilism and Emancipation* that active nihilism is the 'progressive reduction of the original violence of the law' (150). Indeed, the active nihilism of weak ontology 'supplies philosophical reasons for preferring a liberal, tolerant, and democratic society rather than an authoritarian and totalitarian one' (19). And just as Derrida appeals to the literary for its 'revealing power' in the struggle against the 'bad violence' of Western metaphysics, so Vattimo (despite his criticism of deconstruction as a 'pure esthetic game') connects active nihilism with the aesthetic. With this gesture, Vattimo ties weak ontology back to one of the founding works of philosophico-aesthetic modernism, Nietzsche's *Birth of Tragedy* (1872).

As discussed in Chapter 3 above, in *The Birth of Tragedy* Nietzsche locates the birth of modernity in the emergence of 'theoretical man' as the counter-figure to creative/artistic man. Theoretical man, epitomized by Socrates, places all his faith in reason, in the possibility of knowledge, and in truth. These are the values of a modernity that Nietzsche considers to be nihilist in its negation of 'life' as a becoming without purpose, unity, or truth. Vattimo's brand of active nihilism aligns itself with Nietzsche's creative man as a producer of values that are never taken to be anything other than interpretative constructions:

> the interpretations proper to active nihilism are explicitly aware of their own hermeneutic nature and purely on that account correspond to a more adventurous, richer, more open form of life. In the way of life of the herd, in reactive nihilism, no interpretation has the courage to present itself as an interpretation, as someone's interpretation; it must always appear to be objective truth. (Vattimo 2006: 136)

And, crucially, Vattimo follows Nietzsche by connecting his own conception of the active nihilist with the artist:

> if we seek a model for the active nihilist we find Nietzsche directing us to the artist, who is thought of essentially as tragic and Dionysian, but

in a sense that recalls Schopenhauer and his ascetic interpretation of Kantian aesthetic disinterestedness. The possibility of an active nihilism is the capacity, to which the artist bears witness, to transcend the instinct of self-preservation and achieve a condition of moderation that also forms the basis of the heedless, disinterested hubris that is essential for the experimental capacity of the *Übermensch*. (140)

If Vattimo's active nihilist is an artist, then this is because the world that this form of nihilism produces is itself a work of art: 'Once denied any faith in the *Grund* and in the course of events as a development toward an ultimate point, the world appears as a work of art which makes itself' (Vattimo 1988: 96). As for Nietzsche, so for Vattimo, active nihilism is a form of '"aesthetic consciousness"' which 'may be recuperated as an experience of truth precisely insofar as this experience is substantially nihilistic' (114–15).

Returning to our point of departure – Vattimo's claim that active nihilism is 'our (only) chance' in that it works against the violent foundationalism of modernity – we can now say that this hope lies for Vattimo, as for Nietzsche, in the aesthetic. Indeed, Vattimo makes this point explicitly – despite his criticism of deconstruction as a 'pure esthetic game' – when he asserts that the active nihilist 'is able to look at many cultures with a gaze more esthetic than "objective" or truth seeking', and that:

> the salvation of our postmodern civilization can only be an esthetic salvation. [...] the reconciliation of peace and liberty in the postmodern or late-modern world will be attained only on condition that esthetics prevails over objective truth. The variety of lifestyles and the diversity of ethical codes will be able to coexist without bloody clashes only if they are considered as, precisely, styles within an art collection – and for that matter within a museum. (Vattimo 2004: 55–8)

For Vattimo, the active nihilism that alone promises a future freed from the horrors perpetrated by political modernism – from Nazi Germany to Stalinist Russia – is nothing other than 'a general aestheticization of existence' (Vattimo 1988: 52), an endless undoing of foundationalism.

The irony here is that Vattimo's privileging of the aesthetic in the struggle against a certain form of nihilism aligns him with Nietzsche and a tradition of philosophical modernism that includes both Heidegger and Adorno. Although he does not explicitly address this problem of contamination, Vattimo does seek to distinguish his own privileging of the aesthetic from Adorno's. According to Vattimo, Adorno's aesthetics remains an instance of reactive nihilism in that it takes the work of art to be what Vattimo describes as 'a truly prophetic and utopian figuration of an alternative world or of a harmonized existence'. Such an aesthetics is a form of reactive rather than active nihilism in that it conceives of art in terms of founding, 'of figuring *possible* historical worlds which offer an alternative to the existing world' (67; Vattimo's emphasis). Vattimo's own 'general aestheticization of existence' differs from such a foundationalist modernist aesthetics in that it discloses 'the ever new and different determination of regulative structures of experience' (75). The truth disclosed by active nihilism as aesthetic consciousness is truth 'stripped of the authoritarian traits of metaphysical evidence' (76). The difference between Vattimo's conception of the aesthetic and Adorno's is, however, far less clear than Vattimo would have it.

Vattimo's postmodern revalorization of nihilism is, then, a return to Nietzsche, and to philosophical modernism more generally, in two key respects. First, it continues to depend upon an antagonistic relation with a certain form of nihilism, namely reactive nihilism, which is presented as a threat to be countered. Secondly, it locates what Nietzsche terms the 'only superior counterforce' to nihilism in the aesthetic, and more precisely in an aesthetic consciousness that is value-producing rather than value-inheriting.

Aesthetic postmodernism and nihilism

We have seen that the thought of both Derrida (who places deconstruction in antagonistic relation to nihilism) and Vattimo (who revalorizes a certain kind of nihilism as 'our (only) chance' on the grounds that it alone can save us from the terrors of fundamentalism) is marked by two of the most profound traits of philosophical modernism. These are the conception of nihilism as what Derrida terms the 'great question', and the privileging of the aesthetic, in a certain form, as that which can resist the 'bad violence' both of modernity and of the modernist response to modernity. What, though, of aesthetic postmodernism? Is it possible to find therein an escape from the modernity–modernism–nihilism complex? On the one hand, aesthetic postmodernism is like aesthetic modernism in that it has often been charged with nihilism. On the other hand, certain key artists generally seen as postmodern have sought to explore the question of the 'nothing' and of nihilism, without their own work necessarily underwriting nihilism. As for the equivalent within the realm of the aesthetic of Vattimo's philosophical revalorization of nihilism, this too is present, although only rarely in the form of an explicit invocation of nihilism.

One of the most important early theorists of the postmodern in art and literature, Ihab Hassan, aims to distinguish postmodernism from modernism while at the same time making discriminations within aesthetic postmodernism between that which is nihilistic and that which is not. Hassan identifies the composer John Cage as having provided, in a 1952 manifesto, the 'definitive statement on anti-art' in the postmodern period:

> nothing is accomplished by writing a piece of music
> nothing is accomplished by hearing a piece of music
> nothing is accomplished by playing a piece of music
> our ears are now in excellent condition
> (Cage 1968: xii; quoted in Hassan 1975: 22)

Hassan contrasts this statement – which, he claims, reveals a 'sacramental disposition' on Cage's part – with the principles underlying other postmodern aesthetic movements such as neo-Dadaism, pop art, process art, funk art, computer art, concept art, *choseisme*, and *tel quelisme*. These movements Hassan sees as ranging 'from the sacramental to the nihilistic' (22).

Among the various distinctions that Hassan makes between aesthetic modernism and aesthetic postmodernism, one of the most important is that the latter is closer to silence or exhaustion. All forms, all styles, have already been used up, leaving nothing new for the artist to undertake. This might seem to lead inexorably to nihilism, as suggested by the Italian philosopher Massimo Cacciari in his essay 'On the Architecture of Nihilism' (1993). According to Cacciari, what characterizes the 'architecture of nihilism' is that it 'believes every root, form, and traditional symbolic measure to be totally exhausted' (Cacciari 1993: 204). For Hassan, however, this sense of exhaustion does not necessarily lead to nihilism. Indeed, the best postmodern writers 'brilliantly display the resources of the void. Thus the verbal omnipotence of Joyce yields to the omnipotence of Beckett, heir and peer, no less genuine, only more austere' (Hassan 1975: 53). That said, these same artists 'sometimes pass to the other side of silence. The consummation of their art is a work which, remaining art, pretends to abolish itself (Beckett, Tinguely, Robert Morris), or else to become indistinguishable from life (Cage, Rauschenberg, Mailer)' (53). Furthermore, while both modernism and postmodernism are characterized by dehumanization, in postmodern dehumanization irony 'becomes radical, self-consuming play, entropy of meaning' (55). Crucially, however, Hassan aims to distinguish those forms of postmodern art that resist nihilism from those that fall into it. Having described 'nihilism' as a word that 'we often use, when we use it unhistorically, to designate values we dislike', he goes on to claim that when John Cage, in his composition *HPSCHD* (1969), 'insists on Quantity rather than Quality, he does not surrender to nihilism' (53).

Hassan, then, aims to defend aesthetic postmodernism against the charge of nihilism, arguing that the best postmodern art explores silence, exhaustion, and the void without itself being nihilist. A rather different defence of aesthetic postmodernism is offered by Will Slocombe in *Nihilism and the Sublime Postmodern* (2006). Unlike Hassan, and in a manner that owes much to Vattimo, Slocombe distinguishes between two basic kinds of nihilism: modernist and postmodernist. Modernist nihilism 'summarizes those earlier revolutionary formulations of nihilism that sought to escape from certain strictures but which reinforced themselves: authoritarian nihilism'. In contrast, postmodern nihilism 'attempts to escape from even its own strictures. Postmodernism gives rise to the possibility of a self-referential nihilism that is not cultural *ennui* but a liberating (or, at the very least, ethical) turn away from ideological processes: anti-authoritarian nihilism' (Slocombe 2006: 99–100). Slocombe's position is, then, close to Vattimo's in that his distinction between modernist and postmodernist nihilism repeats the Italian philosopher's distinction between reactive and active nihilism, with the latter being championed as a self-reflexive and self-deconstructive anti-foundationalism. Not only does this form of nihilism resist the traditional philosophical foundations such as God, truth, and being, but it also resists the idea of the void as a foundation. Modernist nihilism is 'a totalitarian nothingness where nothing else but nihilism can exist. It is an attempt to remain what Lyotard would term a "good form" by locating itself outside that which it is erasing.

Modernist nihilism is, in effect, a metanarrative, a *weltanschauung* that denies all others.' Postmodern nihilism, on the other hand, 'is concerned with the idea that nihilism cannot truthfully say "there is no truth". This formulation would not attempt to remain outside of that which it negates, meaning that the statement itself would be *both* true *and* untrue, or, as it is nihilism, *neither* true *nor* untrue' (100–1; Slocombe's emphasis). In short, postmodern nihilism includes itself in that which it negates. If Slocombe departs from Vattimo, then he does so by including Nietzsche squarely within modernist nihilism. Whereas Vattimo seeks to demonstrate that Nietzsche's conception of nihilism is already postmodern in its anti-foundationalism, Slocombe claims that Nietzsche 'wrote from "outside" nihilism, rather than from "within" it' (101), and thus failed to include nihilism itself in that which is negated.

Slocombe identifies various examples of postmodernist nihilism in aesthetic postmodernism. In the visual arts, Yves Klein's blue monochromes, Robert Ryman and Robert Rauschenberg's white paintings, Ad Reinhardt's black paintings, and Barnett Newman's works are all 'nihilistic' because they '"make use" of nothingness' (64–5). In the field of music, John Cage's compositions, and above all *4' 33"* (1952), which consists of four minutes and thirty-three seconds of silence, is similarly nihilistic, according to Slocombe. Crucially, this making use of nothingness is not itself simply nothing. The nothingness at the level of the content is countered in the very performativity of the work. John Cage makes this very point when he claims in 'Experimental Music' (1957) that 'There is no such thing as an empty space or an empty time. There is always something to see, something to hear. In fact, try as we may to make silence, we cannot' (Cage 1968: 8).

This conviction that absolute negation is impossible, that neither silence nor the void can be accomplished by the work of art, that neither empty space nor empty time can become an object of experience, lies at the heart of many of the engagements with nihilism in works generally classed as postmodern. A particularly clear example is Donald Barthelme's short story 'Nothing: A Preliminary Account', from the collection *Guilty Pleasures* (1974). In this story, Barthelme treats the question of how one might speak of nothing. In *Watt* (1953), the novel that he wrote while in hiding from the Nazis in the South of France during the Second World War, Samuel Beckett has his narrator, Sam, state that 'the only way one can speak of nothing is to speak of it as though it were something' (Beckett 2009c: 64). In 'Nothing', Barthelme takes a different approach: the only way to speak of nothing is to speak of what it is not. The story begins: 'It's not the yellow curtains. Nor curtain rings' (Barthelme 1993: 245), and the narrator proceeds to list many of the other things that nothingness is not. These include what histories of philosophy often identify as a pre-Socratic form of nihilism. Nothingness, Barthelme's narrator declares, 'is not the nihilism of Gorgias, who asserts that nothing exists and even if something did exist it could not be known and even if it could be known that knowledge could not be communicated, no, it's not that although the tune is quite a pretty one' (246). If nothingness is not nihilism, then that is, of course, because nihilism, whatever form it might take, *is* something, namely a philosophical position.

The unsurprising conclusion reached by the narrator of Barthelme's story is that the list of what nothingness is not 'can in principle never be completed, even if we summon friends or armies to help out [...]. And even if we were able, with much labor, to exhaust the possibilities, get it all *inscribed*, name everything nothing is not, down to the last rogue atom, the one that rolled behind the door, and had thoughtfully included ourselves, the makers of the list, on the list – the list itself would remain' (247; Barthelme's emphasis). Crucially, however, this conclusion leads not to despair but to an attitude akin to the one expressed by Samuel Beckett in *Worstward Ho* (1983), namely 'Fail again. Fail better' (Beckett 1983: 7). Barthelme's story ends: 'What a wonderful list! How joyous that notion that, try as we may, we cannot do other than fail and fail absolutely and that the task will remain always before us, like a meaning for our lives' (Barthelme 1993: 248). Meaning is restored, then, in a task that is strictly speaking endless. Thus, Barthelme's story underwrites the idea of the absurd as articulated by Albert Camus in *The Myth of Sisyphus* (1942), where the affirmation of the absurd is value- and meaning-producing: 'The absurd man says yes and his effort will henceforth be unceasing' (Camus 1975: 110).

A second example of the way in which nihilism becomes the explicit theme of a literary work considered by commentators such as Hassan and Slocombe to be postmodern is John Barth's first novel, *The Floating Opera* (1956). Barth himself describes this work as a 'nihilist comedy', and his next novel, *The End of the Road* (1958), as a 'nihilist catastrophe: the same melody reorchestrated in a grimmer key and sung by a leaner voice' (Barth 1988: vii–viii; quoted in Slocombe 2006: 110). The narrator of *The Floating Opera*, Todd Andrews, wishes to understand why his father hanged himself on Ground-Hog Day, 1930, and in order to do so he undertakes the writing of what he terms his 'Inquiry', based on 'every scrap of information that a human being might gather concerning the circumstances of my father's suicide' (Barth 1981: 222). Just like the attempt to define the nothing through stating what it is not in Barthelme's short story, this 'Inquiry' is 'interminable' (222). And, just as in Barthelme's text, so here, too, this interminability is seen not as a reason for despair but as a valid task. As Barth's narrator puts it: 'It doesn't follow that because a goal is unattainable, one shouldn't work towards its attainment' (223).

Barth's novel engages explicitly with nihilism in chapter 25, in which Todd Andrews sets down as the text of his 'Inquiry' three classically nihilist propositions:

I Nothing has intrinsic value.
II The reasons for which people attribute value to things are always ultimately irrational.
III There is, therefore, no ultimate 'reason' for valuing anything.

(226)

He soon realizes, however, that the 'anything' in his third proposition must include life as such, and so he adds two further propositions:

IV Living is action. There's no final reason for action.
V There's no final reason for living.

(231)

Although the chapter ends with the claim that 'The Inquiry was closed', the narrator later comes to modify his fifth and final proposition to read: 'There's no final reason for living (or for suicide)' (252). According to the literary critic Wayne C. Booth, nihilism and literature are quite simply irreconcilable, since 'To write is to affirm at the very least the superiority of *this* order over *that* order. But superiority according to what code of values? Any answer will necessarily contradict complete nihilism. For the complete nihilist, suicide, not the creation of significant forms, is the only consistent gesture' (Booth 1961: 298; Booth's emphasis). It is precisely this conclusion – that nihilism necessarily leads to suicide – that Barth's novel counters. Taken to its extreme, nihilism strips even suicide of its meaning.

As Barth makes clear in his influential essay 'The Literature of Exhaustion' (1967), the 'exhaustion' that the writer in the postmodern age has to face is not 'the subject of physical, moral, or intellectual decadence', but rather 'the used-upness of certain forms or exhaustion of certain possibilities'. This experience of exhaustion is not, he insists, 'necessarily a cause for despair' (Barth 1977: 70). In short, then, while *The Floating Opera* may thematize nihilism, Barth does not consider the novel itself to be nihilist, and, more generally, does not consider the 'literature of exhaustion' to be nihilist. For an example of such a literature, Barth turns to Jorge Luis Borges's short story 'Pierre Menard, Author of the *Quixote*' (1939), which recounts the seemingly impossible attempt of an imaginary early twentieth-century French writer to compose Cervantes's *Don Quixote*. As Borges's narrator clarifies: 'Pierre Menard did not want to compose *another* Quixote, which surely is easy enough – he wanted to compose *the* Quixote. Nor, surely, need one have to say that his goal was never a mechanical transcription of the original; he had no intention of *copying* it. His admirable ambition was to produce a number of pages which coincided – word for word and line for line – with those of Miguel de Cervantes' (Borges 1999: 91; Borges's emphasis). The narrator observes that this undertaking 'was impossible from the outset' (91), and in this respect it fits perfectly with the impossibility of writing the nothing as articulated in Barthelme's short story. Furthermore, Borges's narrator explicitly presents Menard's efforts as an engagement with nihilism. 'There is no intellectual exercise that is not ultimately pointless', the narrator asserts. Each philosophical doctrine in time becomes no more than 'a mere chapter – if not a paragraph or proper noun – in the history of philosophy'. A similar fate befalls works of literature. In the face of these 'nihilistic observations', Pierre Menard 'resolved to anticipate the vanity that awaits all the labors of mankind; he undertook a task of infinite complexity, a task futile from the outset' (94–5). According to Barth, 'Pierre Menard' is 'a remarkable and original work of literature, the implicit theme of which is the difficulty, perhaps the unnecessity, of writing original works of literature' (Barth 1977: 76). It is also, however, an explicit engagement with the experience of nihilism, and indeed recounts a highly paradoxical attempt to resist nihilism by embracing it: the writer consciously commits himself to a futile undertaking. Again, then, we are returned to the myth of Sisyphus as interpreted by Camus. Accepting the absurd, seemingly committing oneself to nihilism, is presented as a means of countering nihilism.

Like aesthetic modernism, then, aesthetic postmodernism thematizes nihilism, with the difference in treatment lying above all in the attitude or mood taken towards it. Aesthetic postmodernism may seem to revalorize nihilism, but, like the examples of philosophical postmodernism considered above, this revalorization is either merely apparent or else it entails a distinction between forms of nihilism, with the revalorized form being in fact an attempted resistance of nihilism. That the works of Samuel Beckett, like those of the Austrian writer Thomas Bernhard (1931–89), have been taken by some critics to fall within aesthetic modernism and by others to fall within aesthetic postmodernism is just one indication that the distinction between these two forms of the aesthetic is anything but watertight. The difficulty of establishing a clear distinction between aesthetic modernism and aesthetic postmodernism becomes all the more evident when their respective attitudes to nihilism are taken into account. For nihilism can be said to haunt aesthetic postmodernism just as it haunts aesthetic modernism, and to do so in ways that render both the overcoming of nihilism and an identification with it highly problematic.

The spectre of nihilism

Nietzsche's claim, in his autumn 1885–autumn 1886 notebook, that nihilism 'stands at the door' as the 'uncanniest of all guests' has generally been taken to mean that he sees nihilism as being on the point of entering Western thought. As we have seen in Chapter 1, such a reading not only radically simplifies Nietzsche's gestures towards a historiography of nihilism, but also aims to reduce, if not do away with, the uncanniness of nihilism. To begin to appreciate that uncanniness, one has to take account of the ways in which nihilism has haunted philosophical, political, and aesthetic modernism, and continues to haunt philosophical, political, and aesthetic postmodernism, challenging the very distinction between them. The complexities of this haunting become apparent as soon as one starts to consider the ways in which nihilism has been identified as that which both modernism and postmodernism, in their various philosophical, political, and aesthetic forms, seek to counter. In so doing, both modernism and postmodernism tend to privilege the aesthetic, following Nietzsche's claim that the aesthetic is the 'only superior counterforce' to nihilism. On occasion, this counterforce presents itself as a form of nihilism, basing this revalorization on the distinction between forms of nihilism: active and reactive, modernist and postmodernist. It is, however, the sustainability of any such distinction, and above all the sustainability of the distinction between that which is inside and that which is outside nihilism, that is called into question by the very spectrality of nihilism to which Nietzsche points. If nihilism haunts both modernism and postmodernism, then it does so in ways that not only trouble the distinction between them, but also weaken the distinction between nihilism and non-nihilism, nihilism and anti-nihilism – in short, between nihilism and its others.

BIBLIOGRAPHY

Works cited

Ades, Dawn, ed. 2006. *The Dada Reader: A Critical Anthology*. London: Tate Publishing.

Adorno, Theodor W. 1973. *Negative Dialectics*. Trans. E. B. Ashton. London: Routledge & Kegan Paul.

—— 1981. *Prisms*. Trans. Samuel and Shierry Weber. Cambridge, MA: MIT Press.

—— 1982. *Against Epistemology: A Metacritique: Studies in Husserl and the Phenomenological Antinomies*. Trans. Willis Domingo. Oxford: Basil Blackwell.

—— 2000. *Metaphysics: Concept and Problems*. Ed. Rolf Tiedemann. Trans. Edmund Jephcott. Cambridge: Polity Press.

Agamben, Giorgio. 1998. *Homo Sacer: Sovereign Power and Bare Life*. Trans. Daniel Heller-Roazen. Stanford, CA: Stanford University Press.

—— 1999. *Potentialities: Collected Essays in Philosophy*. Ed. and trans. Daniel Heller-Roazen. Stanford, CA: Stanford University Press.

Alexander, Archibald D. B. 1922. *A Short History of Philosophy*. Third edition. Glasgow: Maclehose & Sons.

Anders, Günther. 1960. *Franz Kafka*. Trans. A. Steer and A. K. Thorlby. London: Bowes & Bowes.

Ansell-Pearson, Keith. 1994. *An Introduction to Nietzsche as Political Thinker: The Perfect Nihilist*. Cambridge: Cambridge University Press, 1994.

Badiou, Alain. 2003. *On Beckett*. Ed. Alberto Toscano and Nina Power. Manchester: Clinamen.

—— 2007. *The Century*. Trans. Alberto Toscano. Cambridge: Polity Press.

Bakunin, Michael. 1973. *Selected Writings*. Ed. Arthur Lehning. Trans. Steven Cox and Olive Stevens. London: Jonathan Cape.

Ball, Hugo. 1996. *Flight Out of Time: A Dada Diary*. Ed. John Elderfield. Trans. Ann Raimes. Berkeley, CA: University of California Press.

Barth, John. 1977. 'The Literature of Exhaustion'. In Malcolm Bradbury, ed., *The Novel Today: Contemporary Writers on Modern Fiction*. Manchester: Manchester University Press, 70–83.

—— 1981. *The Floating Opera*. London: Granada.
—— 1988. *The Floating Opera and The End of the Road*. New York: Anchor.
Barthelme, Donald. 1993. *Sixty Stories*. New York: Penguin.
Baudrillard, Jean. 1994. *Simulacra and Simulation*. Trans. Sheila Faria Glaser. Ann Arbor, MI: University of Michigan Press.
Beckett, Samuel. 1938. *Murphy*. London: George Routledge & Sons.
—— 1959. *Molloy, Malone Dies, The Unnamable*. London: John Calder.
—— 1982. *Ill Seen Ill Said*. London: John Calder.
—— 1983. *Worstward Ho*. London: John Calder.
—— 1990. *Complete Dramatic Works*. London: Faber & Faber.
—— 2009a. *The Letters of Samuel Beckett 1929–1940*. Ed. Martha Dow Fehsenfeld and Lois More Overbeck. Cambridge: Cambridge University Press.
—— 2009b. *Molloy*. Ed. Shane Weller. London: Faber & Faber.
—— 2009c. *Watt*. Ed. C. J. Ackerley. London: Faber & Faber.
Benjamin, Walter. 1999a. *The Arcades Project*. Trans. Howard Eiland and Kevin McLaughlin. Cambridge, MA: Belknap Press.
—— 1999b. *Selected Writings*. Vol. 2: *1927–1934*. Ed. Michael W. Jennings, Howard Eiland, and Gary Smith. Trans. Rodney Livingstone et al. Cambridge, MA: Belknap Press.
—— 2002. *Selected Writings*. Vol. 3: *1935–1938*. Ed. Howard Eiland and Michael W. Jennings. Trans. Edmund Jephcott, Howard Eiland et al. Cambridge, MA: Belknap Press.
—— 2003. *Selected Writings*. Vol. 4: *1938–1940*. Ed. Howard Eiland and Michael W. Jennings. Trans. Edmund Jephcott et al. Cambridge, MA: Belknap Press.
Benn, Gottfried. 1970. 'After Nihilism'. *Origin: A Quarterly for the Creative*, series I, nos. 9–12 (1953–54). Nendeln: Kraus Reprint: 97–104.
—— 1976. *Primal Vision: Selected Writings of Gottfried Benn*. Ed. E. B. Ashton. London: Marion Boyars.
—— 1987. *Sämtliche Werke*. Band III. Prosa 1. Ed. Gerhard Schuster. Stuttgart: Klett-Cotta.
—— 2001. *Sämtliche Werke*. Band VI. Prosa 4. Ed. Holger Hof. Stuttgart: Klett-Cotta.
Berlin, Isaiah. 1975. 'Fathers and Children: Turgenev and the Liberal Predicament'. Introduction to Ivan Turgenev, *Fathers and Sons*. Trans. Rosemary Edmonds. Harmondsworth: Penguin, 7–61.
Berman, Marshall. 1983. *All That Is Solid Melts into Air: The Experience of Modernity*. London: Verso.
Biaggi, Vladimir, ed. 1998. *Le Nihilisme*. Paris: GF Flammarion.
Blanchot, Maurice. 1981. *The Madness of the Day / La Folie du jour*. Trans. Lydia Davis. Barrytown, NY: Station Hill.
—— 1982. *The Space of Literature*. Trans. Ann Smock. Lincoln, NE: University of Nebraska Press.

—— 1993. *The Infinite Conversation*. Trans. Susan Hanson. Minneapolis, MN: University of Minnesota Press.

—— 1995a. *The Blanchot Reader*. Ed. Michael Holland. Oxford: Blackwell.

—— 1995b. *The Work of Fire*. Trans. Charlotte Mandell. Stanford, CA: Stanford University Press.

—— 2000. *The Instant of My Death*. Trans. Elizabeth Rottenberg. Stanford, CA: Stanford University Press.

—— 2001. *Faux Pas*. Trans. Charlotte Mandell. Stanford, CA: Stanford University Press.

Booth, Wayne C. 1961. *The Rhetoric of Fiction*. Chicago: University of Chicago Press.

Borges, Jorge Luis. 1999. *Collected Fictions*. Trans. Andrew Hurley. London: Allen Lane.

Bourget, Paul. 1912. *Essais de psychologie contemporaine*. 2 vols. Paris: Plon.

Bradbury, Malcolm, and James McFarlane, eds. 1976. *Modernism: A Guide to European Literature 1890–1930*. Harmondsworth: Penguin.

Brod, Max. 1947. *The Biography of Franz Kafka*. Trans. G. Humphreys Roberts. London: Secker & Warburg.

Büttner, Gottfried. 1984. *Samuel Beckett's Novel 'Watt'*. Trans. Joseph P. Dolan. Philadelphia, PA: University of Pennsylvania Press.

Cacciari, Massimo. 1993. *Architecture and Nihilism: On the Philosophy of Modern Architecture*. Trans. Stephen Sartarelli. New Haven, CT: Yale University Press.

Cage, John. 1968. *Silence: Lectures and Writings*. London: Calder and Boyars.

Calinescu, Matei. 1987. *Five Faces of Modernity: Modernism, Avant-Garde, Decadence, Kitsch, Postmodernism*. Durham, NC: Duke University Press.

Camus, Albert. 1971. *The Rebel*. Trans. Anthony Bower. Harmondsworth: Penguin.

—— 1975. *The Myth of Sisyphus*. Trans. Justin O'Brien. Harmondsworth: Penguin.

Carr, E. H. 1975. *Mikhail Bakunin*. Basingstoke: Macmillan.

Carr, Karen L. 1992. *The Banalization of Nihilism: Twentieth-Century Responses to Meaninglessness*. Albany, NY: State University of New York Press.

Celan, Paul. 1992. *Gesammelte Gedichte in fünf Bänden*. Ed. Beda Allemann and Stefan Reichert. Frankfurt am Main: Suhrkamp.

—— 2002. *Poems*. Trans. Michael Hamburger. Revised and expanded edition. New York: Persea.

—— 2005. *Selections*. Ed. Pierre Joris. Berkeley, CA: University of California Press.

Childs, Peter. 2000. *Modernism*. London: Routledge.

Cioran, E. M. 1970. *The Fall into Time*. Trans. Richard Howard. Chicago: Quadrangle.

—— 1987. *The Temptation to Exist*. Trans. Richard Howard. London: Quartet.

—— 1990. *A Short History of Decay*. Trans. Richard Howard. London: Quartet.
—— 1991. *Anathemas and Admirations*. Trans. Richard Howard. New York: Arcade.
—— 1992. *On the Heights of Despair*. Trans. Ilinca Zarifopol-Johnston. Chicago: University of Chicago Press.
—— 1993. *The Trouble with Being Born*. Trans. Richard Howard. London: Quartet.
—— 1995. *Entretiens*. Paris: Gallimard.
—— 1996. *History and Utopia*. Trans. Richard Howard. London: Quartet.
—— 1999. *All Gall is Divided: Gnomes and Apothegms [Syllogismes de l'amertume]*. Trans. Richard Howard. New York: Arcade.
Clark, Timothy. 1992. *Derrida, Heidegger, Blanchot: Sources of Derrida's Notion and Practice of Literature*. Cambridge: Cambridge University Press.
Corngold, Stanley. 1996. 'Kafka's *The Metamorphosis*: Metamorphosis of the Metaphor'. In Franz Kafka, *The Metamorphosis: Translation, Backgrounds and Contexts, Criticism*. Ed. and trans. Stanley Corngold. New York: W. W. Norton, 79–106.
Crevier, Jean-Baptiste. 1761. *Histoire de l'université de Paris, depuis son origine jusqu'en l'année 1600*. Desaint et Saillant.
Critchley, Simon. 1997. *Very Little … Almost Nothing: Death, Philosophy, Literature*. London: Routledge.
Crosby, Donald A. 1988. *The Specter of the Absurd: Sources and Criticism of Modern Nihilism*. Albany, NY: State University of New York Press.
Darby, Tom, Béla Egyed, and Ben Jones, eds. 1989. *Nietzsche and the Rhetoric of Nihilism: Essays on Intepretation, Language and Politics*. Ottawa: Carleton University Press.
Debord, Guy. 1983. *Society of the Spectacle*. Detroit, MI: Black & Red.
Deleuze, Gilles. 1983. *Nietzsche and Philosophy*. Trans. Hugh Tomlinson. London: Athlone.
—— and Félix Guattari. 1986. *Kafka: Toward a Minor Literature*. Trans. Dana Polan. Minneapolis, MN: University of Minnesota Press.
de Man, Paul. 1986. *The Resistance to Theory*. Ed. Wlad Godzich. Minneapolis, MN: University of Minnesota Press.
—— 1989. *Critical Writings 1953–1978*. Ed. Lindsay Waters. Minneapolis, MN: University of Minnesota Press.
Derrida, Jacques. 1976. *Of Grammatology*. Trans. Gayatri Chakravorty Spivak. Baltimore, MD: Johns Hopkins University Press.
—— 1978. *Writing and Difference*. Trans. Alan Bass. Chicago: Chicago University Press.
—— 1984. 'Deconstruction and the Other'. In Richard Kearney, ed., *Dialogues with Contemporary Thinkers: The Phenomenological Heritage*. Manchester: Manchester University Press, 107–26.

—— 1989. *Memoires: For Paul de Man*. Trans. Cecile Lindsay, Jonathan Culler, Eduardo Cadava, and Peggy Kamuf. Revised edition. New York: Columbia University Press.

—— 1992. *Acts of Literature*. Ed. Derek Attridge. New York: Routledge.

—— 2000. *Demeure: Fiction and Testimony*. Trans. Elizabeth Rottenberg. Stanford, CA: Stanford University Press.

—— 2001. 'I Have a Taste for the Secret'. In Jacques Derrida and Maurizio Ferraris, *A Taste for the Secret*. Ed. Giacomo Donis and David Webb. Trans. Giacomo Donis. Cambridge: Polity Press, 1–92.

—— 2002. *Acts of Religion*. Ed. Gil Anidjar. New York: Routledge.

—— 2005a. *Derrida: Screenplay and Essays on the Film*. Manchester: Manchester University Press.

—— 2005b. *Rogues: Two Essays on Reason*. Trans. Pacale-Anne Brault and Michael Naas. Stanford: Stanford University Press.

—— 2005c. *Sovereignties in Question: The Poetics of Paul Celan*. Ed. Thomas Dutoit and Outi Pasenen. New York: Fordham University Press.

Dostoevsky, Fyodor. 1987. *Selected Letters*. Ed. Joseph Frank and David I. Goldstein. Trans. Andrew R. MacAndrew. New Brunswick: Rutgers University Press.

—— 2000. *Demons*. Trans. Richard Pevear and Larissa Volokhonsky. London: Everyman.

Drieu La Rochelle, Pierre. 1927. 'The Young European'. *transition*, 2 (May): 9–18.

Eaglestone, Robert. 2004. *The Holocaust and the Postmodern*. Oxford: Oxford University Press.

Farías, Victor. 1989. *Heidegger and Nazism*. Ed. Joseph Margolis and Tom Rockmore. Philadelphia, PA: Temple University Press.

Faye, Jean-Pierre, and Michèle Cohen-Halimi. 2008. *L'Histoire cachée du nihilisme: Jacobi, Dostoïevski, Heidegger, Nietzsche*. Paris: La Fabrique.

Feldman, Matthew. 2006. *Beckett's Books. A Cultural History of Samuel Beckett's 'Interwar Notes'*. New York: Continuum.

Felstiner, John. 1995. *Paul Celan: Poet, Survivor, Jew*. New Haven, CT: Yale University Press.

Gibson, Andrew. 2006. *Beckett and Badiou: The Pathos of Intermittency*. Oxford: Oxford University Press.

Gillespie, Michael Allen. 1995. *Nihilism before Nietzsche*. Chicago: University of Chicago Press.

Glicksberg, Charles I. 1975. *The Literature of Nihilism*. London: Associated University Presses.

Graver, Lawrence, and Raymond Federman, eds. 1979. *Samuel Beckett: The Critical Heritage*. London: Routledge & Kegan Paul.

Griffin, Roger. 2007. *Modernism and Fascism: The Sense of a Beginning under Mussolini and Hitler*. Basingstoke: Palgrave Macmillan.

Habermas, Jürgen. 1987. *The Philosophical Discourse of Modernity.* Cambridge, MA: MIT Press.

Hamburger, Michael. 1954. 'Art and Nihilism: The Poetry of Gottfried Benn'. *Encounter*, 3(4): 49–59.

Harrigan, Anthony. 1998. 'Post-Modern Nihilism in America'. *St. Croix Review*, 31(5): 24–32.

Harvey, David. 1989. *The Condition of Postmodernity: An Enquiry into the Origins of Cultural Change.* Oxford: Basil Blackwell.

Hassan, Ihab. 1975. *Paracriticisms: Seven Speculations on the Times.* Urbana, IL: University of Illinois Press.

Heidegger, Martin. 1962. *Being and Time.* Trans. John Macquarrie and Edward Robinson. New York: Harper & Row.

—— 1968. *What Is Called Thinking?.* Trans. J. Glenn Gray. New York: Harper & Row.

—— 1971a. *On the Way to Language.* Trans. Peter D. Hertz. New York: Harper & Row.

—— 1971b. *Poetry, Language, Thought.* Trans. Albert Hofstadter. New York: Harper & Row.

—— 1977. *The Question Concerning Technology and Other Essays.* Trans. William Lovitt. New York: Harper & Row.

—— 1979. *Nietzsche.* Vol. 1: *The Will to Power as Art.* Trans. David Farrell Krell. New York: Harper & Row.

—— 1982a. *Hölderlins Hymne 'Andenken'.* Frankfurt am Main: Vittorio Klostermann.

—— 1982b. *Nietzsche.* Vol. 4: *Nihilism.* Ed. David Farrell Krell. Trans. Frank A. Capuzzi. New York: Harper & Row.

—— 1989. *Hölderlins Hymnen 'Germanien' und 'Der Rhein'.* Frankfurt am Main: Vittorio Klostermann.

—— 1998. *Pathmarks.* Ed. William McNeill. Cambridge: Cambridge University Press.

—— 1999. *Contributions to Philosophy (From Enowning).* Trans. Parvis Emad and Kenneth Maly. Bloomington, IN: Indiana University Press.

—— 2000a. *Elucidations of Hölderlin's Poetry.* Trans. Keith Hoeller. Amherst, NY: Humanity Books.

—— 2000b. *Introduction to Metaphysics.* Trans. Gregory Fried and Richard Polt. New Haven, CT: Yale University Press.

—— 2002. *Off the Beaten Track [Holzwege].* Ed. and trans. Julian Young and Kenneth Haynes. Cambridge: Cambridge University Press.

—— 2003a. *Four Seminars: Le Thor 1966, 1968, 1969, Zähringen 1973.* Trans. Andrew Mitchell and François Raffoul. Bloomington, IN: Indiana University Press.

—— 2003b. *Philosophical and Political Writings.* Ed. Manfred Stassen. New York: Continuum.

Heller, Peter. 1966. *Dialectics and Nihilism: Essays on Lessing, Mann and Kafka*. Amherst, MA: University of Massachusetts Press.

Hemingway, Ernest. 1934. *Winner Take Nothing*. London: Jonathan Cape.

Hill, Leslie. 2001. *Bataille, Klossowski, Blanchot: Writing at the Limit*. Oxford: Oxford University Press.

Hillis Miller, J. 1979. 'The Critic as Host'. In Harold Bloom et al., *Deconstruction and Criticism*. London: Routledge & Kegan Paul, 217–53.

Hingley, Ronald. 1967. *Nihilists: Russian Radicals and Revolutionaries in the Reign of Alexander II (1855–81)*. New York: Delacorte Press.

Hölderlin, Friedrich. 1951. *Sämtliche Werke*. Vol. II: *Gedichte nach 1800*. Ed. Friedrich Beissner. Stuttgart: W. Kohlhammer.

Hopkins, David. 2004. *Dada and Surrealism: A Very Short Introduction*. Oxford: Oxford University Press.

Horkheimer, Max, and Theodor W. Adorno. [1947] 2002. *Dialectic of Enlightenment*. Trans. John Cumming. New York: Continuum.

Huelsenbeck, Richard, ed. 1993. *The Dada Almanac* [Berlin, 1920]. London: Atlas.

Hugo, Victor. 1907. *Les Misérables*. In *Works of Victor Hugo*. Vol. 2. New York: The Nottingham Society.

Jacobi, Friedrich Heinrich. 1987. 'Open Letter to Fichte'. In Ernst Behler, ed., *Philosophy of German Idealism*. New York: Continuum, 119–41.

Jameson, Fredric. 1994. *The Seeds of Time*. The Wellek Library Lectures at the University of California, Irvine. New York: Columbia University Press.

—— 1998. *The Cultural Turn: Selected Writings on the Postmodern, 1983–1998*. London: Verso.

Jolas, Eugene. 1927. 'Gottfried Benn'. *transition*, 5 (August): 146–9.

Juliet, Charles. 1995. *Conversations with Samuel Beckett and Bram Van Velde*. Trans. Janey Tucker.

Jünger, Ernst. 1980. *Sämtliche Werke*. Zweite Abteilung: Essays, Band 7, Essays I: *Betrachtungen zur Zeit*. Stuttgart: Klett-Cotta.

—— 1981. *Sämtliche Werke*. Zweite Abteilung: Essays, Band 8, Essays II: *Der Arbeiter*. Stuttgart: Klett-Cotta.

—— 1993. 'Total Mobilization'. Trans. Joel Golb and Richard Wolin. In Richard Wolin, ed., *The Heidegger Controversy: A Critical Reader*. Cambridge, MA: MIT Press, 122–39.

Kafka, Franz. 1973. *Shorter Works*. Ed. and trans. Malcolm Pasley. London: Secker & Warburg.

—— 1990. *Tagebücher*. Kritische Ausgabe. Ed. Hans-Gerd Koch, Michael Müller, and Malcolm Pasley. Frankfurt am Main: S. Fischer.

—— 1991. *The Blue Octavo Notebooks*. Ed. Max Brod. Trans. Ernst Kaiser and Eithne Wilkins. Cambridge, MA: Exact Exchange.

—— 1992a. *Nachgelassene Schriften und Fragmente*. Vol. II. Ed. Jost Schillemeit. Frankfurt am Main: S. Fischer.

—— 1992b. *The Trial*. Trans. Willa and Edwin Muir. Revised by E. M. Butler. New York: Alfred A. Knopf.

—— 1993. *Collected Stories*. Ed. Gabriel Josipovici. New York: Alfred A. Knopf.

—— 1994. *Drucke zu Lebzeiten*. Ed. Hans-Gerd Koch, Wolf Kittler and Gerhard Neumann. Frankfurt am Main: S. Fischer.

Kermode, Frank. 1967. *The Sense of an Ending: Studies in the Theory of Fiction*. New York: Oxford University Press.

Kuhn, Elisabeth. 1984. 'Nietzsches Quelle des Nihilismus-Begriffs'. *Nietzsche-Studien*, 13: 253–78.

Laclau, Ernesto. 1988. 'Politics and the Limits of Modernity'. In Andrew Ross, ed., *Universal Abandon? The Politics of Postmodernism*. Edinburgh: Edinburgh University Press, 63–82.

Lacoue-Labarthe, Philippe. 1990. *Heidegger, Art and Politics: The Fiction of the Political* [*La Fiction du politique*]. Trans. Chris Turner. Oxford: Basil Blackwell.

—— 2007. *Heidegger and the Politics of Poetry*. Trans. Jeff Fort. Urbana and Chicago: University of Illinois Press.

Lawrence, D. H. 1987. *Women in Love*. Ed. David Farmer, Lindeth Vasey, and John Worthen. Cambridge: Cambridge University Press.

Levinas, Emmanuel. 1996. *On Maurice Blanchot*. In *Proper Names*. Trans. Michael B. Smith. London: Athlone, 125–87.

Lewis, Wyndham. 1952. *The Writer and the Absolute*. London: Methuen.

—— 1994. 'The Diabolical Principle'. In *The Enemy: A Review of Art and Literature, Number 3 (1929)*. Ed. David Peters Corbett. Santa Rosa, CA: Black Sparrow Press, 9–84.

Lippard, Lucy R., ed. 1971. *Dadas on Art*. Englewood Cliffs, NJ: Prentice-Hall.

Loose, Gerhard. 1974. *Ernst Jünger*. New York: Twayne.

Löwith, Karl. 1995. *Martin Heidegger and European Nihilism*. Ed. Richard Wolin. Trans. Gary Steiner. New York: Columbia University Press.

Lukács, Georg. 1963. *The Meaning of Contemporary Realism* [*Wider den missverstandenen Realismus*]. Trans. John and Necke Mander. London: Merlin.

Lyotard, Jean-François. 1984. *The Postmodern Condition: A Report on Knowledge*. Trans. Geoff Bennington and Brian Massumi. Manchester: Manchester University Press.

Musil, Robert. 1952. *Der Mann ohne Eigenschaften*. Hamburg: Rowohlt.

Nevin, Thomas. 1997. *Ernst Jünger and Germany: Into the Abyss, 1914–1945*. London: Constable.

Nietzsche, Friedrich. [1872] 1967. *The Birth of Tragedy and The Case of Wagner*. Trans. Walter Kaufmann. New York: Vintage.

—— [1901, 1906] 1968. *The Will to Power*. Ed. Walter Kaufmann. Trans. Walter Kaufmann and R. J. Hollingdale. New York: Vintage.

—— 1969. *Selected Letters*. Ed. and trans. Christopher Middleton. Chicago and London: University of Chicago Press.

—— 1974. *The Gay Science, with a Prelude in Rhymes and an Appendix of Songs*. Trans. Walter Kaufmann. New York: Vintage.

—— 1989. *On the Genealogy of Morals*. Trans. Walter Kaufmann and R. J. Hollingdale. In *On the Genealogy of Morals / Ecce Homo*. Ed. Walter Kaufmann. New York: Vintage.

—— 1997. *Untimely Meditations*. Ed. Daniel Breazeale. Trans. R. J. Hollingdale. Cambridge: Cambridge University Press.

—— 1999a. *The Birth of Tragedy and Other Writings*. Ed. Raymond Geuss and Ronald Speirs. Trans. Ronald Speirs. Cambridge: Cambridge University Press.

—— 1999b. *Sämtliche Werke*. Kritische Studienausgabe. Ed. Giorgio Colli and Mazzino Montinari. 15 vols. Berlin: de Gruyter.

—— 2002. *Beyond Good and Evil: Prelude to a Philosophy of the Future*. Ed. Rolf-Peter Horstmann and Judith Norman. Trans. Judith Norman. Cambridge: Cambridge University Press.

—— 2003. *Writings from the Late Notebooks*. Ed. Rüdiger Bittner. Trans. Kate Sturge. Cambridge: Cambridge University Press.

—— 2005. *The Anti-Christ, Ecce Homo, Twilight of the Idols and Other Writings*. Ed. Aaron Ridley and Judith Norman. Trans. Judith Norman. Cambridge: Cambridge University Press.

—— 2006. *Thus Spoke Zarathustra: A Book for All and None*. Ed. Adrian Del Caro and Robert B. Pippin. Trans. Adrian Del Caro. Cambridge: Cambridge University Press.

Osborne, Peter. 1996. *The Politics of Time: Modernity and Avant-Garde*. London: Verso.

Paul, Elliot. 1927. 'The New Nihilism'. *transition*, 2 (May): 164–8.

Podhoretz, Norman. 1958. 'The New Nihilism'. *Partisan Review*, 15 (Fall): 576–90.

Pöggeler, Otto. 1993. 'Heidegger's Political Self-Understanding'. Trans. Steven Galt Crowell. In Richard Wolin, ed., *The Heidegger Controversy: A Critical Reader*. Cambridge, MA: MIT Press, 198–244.

Poggioli, Renato. 1968. *The Theory of the Avant-Garde*. Trans. Gerald Fitzgerald. Cambridge, MA: Belknap Press.

Rauschning, Hermann. 1939. *Germany's Revolution of Destruction* [*Die Revolution des Nihilismus*]. Trans. E. W. Dickes. London: William Heinemann.

Riha, Karl, and Jörgen Schäfer, eds. 1994. *DADA total. Manifeste, Aktionen, Texte, Bilder*. Stuttgart: Philipp Reclam.

Rochefort, Robert. 1955. *Kafka oder die unzerstörbare Hoffnung*. Vienna: Herold.

Safranski, Rüdiger. 1998. *Martin Heidegger: Between Good and Evil*. Trans. Ewald Osers. Cambridge, MA: Harvard University Press.

Schaffner, Anna Katharina. 2007. *Sprachzerlegung in historischer Avantgardelyrik und konkreter Poesie*. Berlin: ECA.

—— 2010. 'Kafka and the Hermeneutics of Sadomasochism'. *Forum for Modern Language Studies*, 46(3): 334–50.

Scholem, Gershom. 1941. *Major Trends in Jewish Mysticism: The Hilda Stich Strook Lectures, 1938*. Jerusalem: Schocken.

—— 1970. 'Zehn unhistorische Sätze über Kabbala'. *Judaica*, 3: 264–71.

—— ed. 1989. *The Correspondence of Walter Benjamin and Gershom Scholem 1932–1940*. Trans. Gary Smith and Andre Lefevere. New York: Schocken Books.

Schopenhauer, Arthur. 1969. *The World as Will and Representation*. Trans. E. F. J. Payne. 2 vols. New York: Dover.

Sheppard, Richard. 2000. *Modernism–Dada–Postmodernism*. Evanston, IL: Northwestern University Press.

Slocombe, Will. 2006. *Nihilism and the Sublime Postmodern: The (Hi)Story of a Difficult Relationship from Romanticism to Postmodernism*. New York: Routledge, 2006.

Spengler, Oswald. 1932. *The Decline of the West*. Trans. Charles Francis Atkinson. London: Allen & Unwin.

Szondi, Peter. 2003. *Celan Studies*. Ed. Jean Bollack et al. Trans. Susan Bernofsky and Harvey Mendelsohn. Stanford, CA: Stanford University Press.

Trakl, Georg. 2001. *Poems and Prose*. Trans. Alexander Stillmark. London: Libris.

Turgenev, Ivan. 1975. *Fathers and Sons*. Trans. Rosemary Edmonds. Harmondsworth: Penguin.

Tzara, Tristan. 1992. *Seven Dada Manifestos and Lampisteries*. Trans. Barbara Wright. London: Calder.

Vaneigem, Raoul. 1983. *The Revolution of Everyday Life* [*Traité de savoir-vivre à l'usage des jeunes générations*]. Trans. Donald Nicholson-Smith. London: Aldgate Press.

Vattimo, Gianni. 1988. *The End of Modernity: Nihilism and Hermeneutics in Post-Modern Culture*. Trans. Jon R. Snyder. Cambridge: Polity Press.

—— 1998. 'The Trace of the Trace'. In Jacques Derrida and Gianni Vattimo, eds, *Religion*. Cambridge: Polity Press, 79–94.

—— 2004. *Nihilism and Emancipation: Ethics, Politics, Law*. Ed. Santiago Zabala. Trans. William McCuaig. New York: Columbia University Press.

—— 2006. *Dialogue with Nietzsche*. Trans. William McCuaig. New York: Columbia University Press.

Virilio, Paul. 1991. *The Aesthetics of Disappearance*. Trans. Philip Beitchman. New York: Semiotext(e).

Weber, Alfred. 1947. *Farewell to European History, or The Conquest of Nihilism*. Trans. R. F. C. Hull. London: Kegan Paul, Trench, Trubner & Co.

Wellek, René. 1990. 'The New Nihilism in Literary Studies'. In François Jost and Melvin J. Friedman, eds, *Aesthetics and the Literature of Ideas: Essays in Honor of A. Owen Aldridge*. Newark: University of Delaware Press, 77–85.

Weller, Shane. 2008. *Literature, Philosophy, Nihilism: The Uncanniest of Guests.* Basingstoke: Palgrave Macmillan.

Wolin, Richard, ed. 1993. *The Heidegger Controversy: A Critical Reader.* Cambridge, MA: MIT Press.

Wolosky, Shira. 1995. *Language Mysticism: The Negative Way of Language in Eliot, Beckett, and Celan.* Stanford, CA: Stanford University Press.

Wurm, Franz. 1995. 'Erinnerung'. In Paul Celan and Franz Wurm, *Briefwechsel*. Ed. Barbara Wiedemann and Franz Wurm. Frankfurt am Main: Suhrkamp, 245–51.

Zola, Emile. 1923. *Les Romanciers naturalistes.* Paris: Fasquelle.

Further reading

[In addition to the above list of works cited, the following works are recommended further reading on the topic of modernism and nihilism.]

Adams, Robert Martin. 1966. *Nil: Episodes in the Literary Conquest of the Void during the Nineteenth Century.* London: Oxford University Press.

Ansell-Pearson, Keith, and Diane Morgan, eds. 2000. *Nihilism Now! Monsters of Energy.* Basingstoke: Macmillan.

Cunningham, Conor. 2002. *Genealogy of Nihilism: Philosophies of Nothing and the Difference of Theology.* London: Routledge.

Diken, Bülent. 2009. *Nihilism.* New York and London: Routledge.

Dryzhakova, Elena. 1980. 'Dostoyevsky, Chernyshevsky, and the Rejection of Nihilism'. *Oxford Slavonic Papers*, 13: 58–79.

Goudsblom, Johan. 1980. *Nihilism and Culture.* Oxford: Blackwell.

Kay, Wallace G. 1971. 'Blake, Baudelaire, Beckett: The Romantics of Nihilism'. *Southern Quarterly*, 9: 253–9.

King, Anthony. 1998. 'Baudrillard's Nihilism and the End of Theory'. *Telos*, 112: 89–106.

Levin, David Michael. 1988. *The Opening of Vision: Nihilism and the Postmodern Situation.* London: Routledge.

Paterson, R. W. K. 1971. *The Nihilistic Egoist: Max Stirner.* Oxford: Oxford University Press.

Rose, Gillian. 1984. *Dialectic of Nihilism: Post-Structuralism and Law.* Oxford: Basil Blackwell.

Rosen, Stanley. 1969. *Nihilism: A Philosophical Essay.* New Haven, CT: Yale University Press.

Stepniak. 1883. *Underground Russia: Revolutionary Profiles and Sketches from Life.* Translated from the Italian. London: Smith, Elder, & Co.

INDEX

absurd, 11, 65–6, 81–2, 161, 163–4
Adorno, Theodor W., 2–4, 7–9, 14, 19–20, 40, 60–7, 69, 116–19, 124, 126–7, 133, 143–4, 147, 156–7
Aeschylus, 81
aestheticism, 4
Agamben, Giorgio, 148–50
Alexander, Archibald, 128
Alexander II, Tsar, 23
alterity, 20, 61, 63, 143–7, 150, 155
Amiel, Henri-Frédéric, 78
amoralism, 103
anarchism, 23, 33, 37, 97–8
Anders, Günther, 115
Andreyev, Leonid, 115
Ansell-Pearson, Keith, 26–7, 38
anti-humanism, 10, 14
anti-Semitism, 8, 21, 25, 46–7, 49, 60–1, 67, 149
Antisthenes, 43
Antonioni, Michelangelo, 13
Artaud, Antonin, 109, 146
atheism, 9, 14, 17–18, 20–1
Attridge, Derek, 147
Auschwitz (*see also* Holocaust), 3, 61–3, 133
avant-garde, 4, 9, 14, 77, 92–102, 150

Baader, Franz von, 98
Badiou, Alain, 101, 127
Baeumler, Alfred, 42, 49
Bakunin, Mikhail, 23–4, 36, 97–8
Ball, Hugo, 7, 14, 94, 97–101
Balzac, Honoré de, 66, 79, 105, 124
Barrès, Maurice, 47

Barth, John, 161–3
Barthelme, Donald, 160–1, 163
Baudelaire, Charles, 5, 28, 77–9, 107
Baudrillard, Jean, 14, 150–1
Beckett, Samuel, 64–5, 72, 77, 102, 113–15, 125–30, 133, 135–6, 159–61, 164
Beethoven, Ludwig van, 84
Benjamin, Walter, 5, 9, 93, 106, 111–13, 115–16, 135, 144, 147–9
Benn, Gottfried, 6–7, 47, 104–6, 109, 113–14
Berlin, Isaiah, 22
Berman, Marshal, 114
Bernanos, Georges, 47
Bernhard, Thomas, 164
biopolitics, 18, 148–9
Blanchot, Maurice, 7, 9, 14, 68–70, 73, 116–18, 124, 130–3, 146
bolshevism, 9
Booth, Wayne C., 130, 162
Borges, Jorge Luis, 163–4
Boundas, Constantin, 144
Bourget, Paul, 9, 28, 78–80
Bradbury, Malcolm, 9
Breton, André, 92, 94
Brod, Max, 111–13, 119–21, 123
Buddhism, 29, 36, 40, 43, 80
Bürger, Peter, 95
Büttner, Gottfried, 127

Cacciari, Massimo, 158
Cage, John, 158–60
Calinescu, Matei, 43, 78
Camus, Albert, 7, 9, 11, 14, 65–70, 108–9, 115, 125, 161, 163

capitalism, 2, 45, 52, 60, 104, 114
Carnot, Marie-François-Sadi, 23
Carr, Karen L., 11–12, 141–2
Celan, Paul, 77, 133–6, 146
Céline, Louis-Ferdinand, 47, 79, 109, 113
Cervantes, Miguel de, 79, 163
Childs, Peter, 9
Christianity, 9, 14, 29–32, 34–5, 37, 39–40, 46, 61, 80, 83–5, 98, 100, 103, 124, 140, 149
Cioran, E. M., 7, 9, 14, 70–3
Clark, Timothy, 145
Cloots, Anarcharsis, 17–19, 21, 139
Commune, Paris, 78
communism, 5, 52, 60, 67, 104
Conrad, Joseph, 47
Corngold, Stanley, 123
Crevier, Jean-Baptiste-Louis, 18–19, 21
Critchley, Simon, 127
Crosby, Donald A., 10–12

Dada, 9, 77, 92–102, 112, 115, 139, 150, 158
Darwin, Charles, 105
'death of God', 30, 100–1, 110, 114, 116, 150, 152
Debord, Guy, 7
decadence, 1, 26, 29, 33, 70–1, 78, 101–2, 104, 150, 163
deconstruction, 9, 14, 142–7, 150, 154–6, 159
Deleuze, Gilles, 32–3, 123, 154
de Man, Paul, 142–3, 147–9
Democritus, 128–9, 135
Derrida, Jacques, 14, 18, 127, 135–6, 142–7, 150, 152, 154–5, 157
desacralization, 2, 4–5
Descartes, René, 55, 59
Dichtung, 8, 48, 58–9, 86–92, 146
Dickens, Charles, 79
Dionysius the Areopagite, 101

Dostoevsky, Fyodor, 24–6, 28–9, 47, 66–8, 72, 103, 107, 114–15
Drieu La Rochelle, Pierre, 8, 103–4, 107, 139
Duchamp, Marcel, 9, 14, 92, 102
Dumas *fils*, Alexandre, 78
Dürer, Albrecht, 105

Eaglestone, Robert, 4
Einhorn, Erich, 135
Eliot, T. S., 109
Elisabeth ('Sisi'), Empress of Austria, 23
Enlightenment, 2–3, 21, 31–3, 35, 40, 44, 70, 72, 124, 140–1, 154
Epicurus, 33, 43
eternal recurrence, 36–8, 66
Expressionism, 92, 113

fascism (*see also* Nazism), 1, 3–5, 7, 9, 14, 18, 49, 64, 71, 104, 113
Faulkner, William, 47, 113–14
Faye, Jean-Pierre, 9, 29
Felstiner, John, 135
Ferraris, Maurizio, 144
Fichte, Johann Gottlieb, 19–21, 25
First World War, 4, 7, 25, 27, 43, 45, 89, 102–3
Flaubert, Gustave, 5, 28, 77–80
Franco-Prussian War, 4, 78
Frank, Joseph, 29
Frankfurt School, 2
French Revolution, 3, 9, 17–19, 24, 31, 43–6, 124, 139
Futurism, 92–3

Gautier, Théophile, 78
Genet, Jean, 125, 146
George, Stefan, 86
Gibson, Andrew, 17
Gide, André, 79, 108
Gillespie, Michael Allen, 36
Glicksberg, Charles I., 115, 119, 125

Gnosticism, 126
Goethe, Johann Wolfgang von, 43, 105
Goncourt, Brothers, 78
Gorgias, 128, 161
Greene, Graham, 47
Griffin, Roger, 1–7, 9, 12, 18, 26, 31–3, 35, 37, 79–80, 123
Grossman, Evelyne, 145
Guattari, Félix, 123
Gutwirth, Marcel, 130

Habermas, Jürgen, 58
Hamburger, Michael, 105
Harrigan, Andrew, 140
Harvey, David, 1
Hassan, Ihab, 158–9, 161
Hegel, G. W. F., 23, 38, 58, 63, 118, 131
Heidegger, Martin, 3–4, 6–9, 14, 19–20, 24, 26–7, 32–3, 40, 42, 48–60, 62, 64–5, 68–9, 71, 73, 86–92, 108, 117, 132, 141, 143, 146–7, 151–2, 156
Heine, Heinrich, 141
Heller, Erich, 147
Heller, Peter, 114–15, 118–19, 121–3
Hellingrath, Norbert von, 86
Hemingway, Ernest, 47, 102, 108, 129
Heydrich, Reinhard, 47
Hill, Leslie, 68–9
Hingley, Ronald, 24
Hitler, Adolf, 6, 9, 46, 50, 53, 60, 106
Hölderlin, Friedrich, 8, 58–9, 65, 86–9, 91, 117, 147
Holocaust (*see also* Auschwitz), 3–4, 48, 61–3, 67, 89–90, 118, 133, 135
Homer, 85, 91
Hopkins, David, 94
Horkheimer, Max, 2–4, 66
Howe, Mary Manning, 128
Huelsenbeck, Richard, 95, 97, 100, 112
Hugo, Victor, 21

humanism, 9, 14
Husserl, Edmund, 108
Huxley, Aldous, 79

Ibsen, Henrik, 43
industrial revolution, 31
Ionesco, Eugene, 115

Jacobi, Friedrich Heinrich, 19–21, 25
Jameson, Fredric, 2, 5
Janko, Marcel, 92–3, 97
Jarry, Alfred, 125
Jolas, Eugene, 104, 106–7
Joyce, James, 79, 102, 107, 113–14, 146, 158
Juliet, Charles, 127
Jung, Carl Gustav, 113
Jung, Franz, 113
Jünger, Ernst, 6–9, 14, 42–8, 53, 57–8, 64, 69, 71, 73, 103, 132

Kafka, Franz, 6, 9, 12, 14, 47, 64–6, 77, 94, 101–2, 109–27, 131–6, 146, 148–9
Kandinsky, Wassily, 100–1
Kant, Immanuel, 3, 19–21, 31, 140, 156
Katkov, M. N., 25
Kaun, Axel, 128
Kazantzakis, Nikos, 115
Kearney, Richard, 142
Kermode, Frank, 6, 8
Kierkegaard, Søren, 65, 115
Klee, Paul, 108
Klein, Yves, 160
Krieck, Ernst, 50
Kuhn, Elisabeth, 28

Laclau, Ernesto, 140
Lacoue-Labarthe, Philippe, 8, 59, 62, 88
Lagarde, Paul, 53
Langbehn, Julius, 53

Lautréamont, 47, 107, 125
Lawrence, D. H., 14, 79, 109
Leconte de Lisle, Charles-Marie-René, 78
Levinas, Emmanuel, 130–1
Lewis, Wyndham, 7–8, 14, 106–7, 109
Lombard, Peter, 19, 26
London, Jack, 10
Loose, Gerhard, 45–6
Löwith, Karl, 79
Lukács, Georg, 8–9, 14, 108, 110–11, 113–15, 124–5, 134–5
Lyotard, Jean-François, 3, 140, 159

Maikov, A. N., 25
Mailer, Norman, 159
Mallarmé, Stéphane, 145–6
Malraux, André, 47, 66, 79, 108–9, 115
Mann, Thomas, 79, 109, 124
Marx, Karl, 25, 43, 53, 114, 140
Maupassant, Guy de, 22, 28, 78
Mauthner, Fritz, 10–11
McFarlane, James, 9
Melville, Herman, 66
Mérimée, Prosper, 28, 78
Meshchersky, V. P., 25
metapolitics, 4
Miller, Henry, 48, 109
Miller, J. Hillis, 142
Montherlant, Henry de, 47, 115
Morris, Robert, 159
Musil, Robert, 109, 113–14
Mussolini, Benito, 53
mysticism (Jewish), 110–11
myth, 7–8, 14, 58, 83–4, 140

Nadeau, Maurice, 125
Nazism, 3–4, 6, 9, 26, 32, 42, 49–50, 53–5, 59–62, 67, 86, 106, 118, 143–4, 146, 148–9, 156, 160
Nechaev, Sergei, 14, 24–5, 36, 67, 139

Newman, Barnett, 160
Nietzsche, Friedrich, 3–4, 6–7, 9–14, 18, 26–43, 45, 47–55, 57–9, 62–71, 77, 80–8, 91, 94, 96–8, 100–1, 103–5, 107–8, 110, 114–15, 117–18, 123–4, 133, 136, 140, 143, 147, 149–57, 160, 164–5
nihilism, forms of,
    active, 18, 35–8, 45–6, 57, 67, 90, 94–5, 106, 151–7, 159, 165
    aesthetic, 19, 104, 106
    aletheiological, 11, 130
    anthropological, 113
    axiological, 11
    cosmological, 10–11
    epistemological, 10–11
    ethical, 11, 130
    existential, 10–11, 108
    imperfect, 149–50
    individual, 67
    mediatic, 151–2
    medical, 113
    messianic, 149
    metaphysical, 11–12
    moral, 10–11
    mystical, 112
    'new', 102–4, 109–10, 139
    ontological, 11, 130
    passive, 35–6, 39, 57, 67, 94, 124
    perfect, 34, 37, 149–50
    political, 10–11, 17–19, 21–7, 36–7, 148–9
    reactive, 35, 106, 152–3, 155, 157, 159, 165
    Russian, 10, 14, 21–6, 36–7, 67, 97–8, 103, 139
    state, 67
    theological, 18–22
    uncanny, 13–14, 34, 38, 40–1, 48, 51, 62, 69, 77, 133, 136, 164–5

Osborne, Peter, 43
Overbeck, Franz, 28, 85

palingenesis, 1
Paul, Elliot, 102–4, 106–7, 109, 139
pessimism, 28, 31–3, 78, 94, 106, 115, 117–18
Picabia, Francis, 94–6
Picasso, Pablo, 108
Pindar, 91
Pingaud, Bernard, 125
Pisarev, Dimitri, 14, 98, 139
Plato, 32–4, 37, 52, 55, 85
Plutarch, 129
Podhoretz, Norman, 109–10
Poggioli, Renato, 93
postmodernism, 4, 12–13, 64, 69, 136, 139–65
postmodernity, 3–4, 10, 13, 39
Pound, Ezra, 8
progress, concept of, 2, 4, 8, 31, 33, 45–6, 70, 118, 124, 141, 154
Protagoras, 100
Proust, Marcel, 47, 66, 79
Putsykovich, V. F., 25
Pyrrho, 33

Rauschenberg, Robert, 159–60
Rauschning, Hermann, 61
reification, 63
Reinhardt, Ad, 160
Renan, Ernest, 78
Rilke, Rainer Maria, 47, 86, 89
Rimbaud, Arthur, 47, 107
Rochefort, Robert, 121–2
Robbe-Grillet, Alain, 110
Robespierre, Maximilien, 18
romanticism, 4, 32, 45, 48, 58, 78–80, 84, 86, 91, 107, 147, 150
Rorty, Richard, 154
Rosenberg, Alfred, 46
Russell, Bertrand, 10–11

Russian nihilism: *see* nihilism, Russian
Russian Revolution, 104
Ryman, Robert, 160

Sabbatianism, 112
Sade, Marquis de, 3, 66, 107, 125
Sappho, 91
Sarraute, Nathalie, 109–10
Sartre, Jean-Paul, 108, 110, 115, 125
Saussure, Ferdinand de, 99
scepticism, 72, 143
Schaffner, Anna Katharina, 120, 122
Schelling, F. W. J., 53
Scholem, Gershom, 111–13, 135, 148–9
Schopenhauer, Arthur, 11, 28, 33, 36, 38, 81, 83, 124, 156
science, 2, 20, 30–1, 33, 40, 50, 78–9, 83–5, 114
Second World War, 3, 27, 48, 55, 59–61, 70, 89, 91, 106–9, 125, 132, 143, 152, 160
secularization, 2
Shakespeare, William, 11, 32
Sheppard, Richard, 94
Simon, Claude, 110
Situationists, 93–4
Slocombe, Will, 159–61
socialism, 9, 14, 23–6, 36
Socrates, 43, 83, 155
Sontag, Susan, 70
Sophocles, 81, 87, 91
Spengler, Oswald, 9, 42–3, 47
Stalinism, 61, 156
Stein, Gertrude, 102, 107
Stendhal, 28, 66, 78
Stillmark, Alexander, 90
Stirner, Max, 10
Stoicism, 43
Strauss, Leo, 61
Surrealism, 92, 150
Szondi, Peter, 134

Taine, Hyppolite, 78
technology, 2, 9, 45–6, 48, 52–3, 58, 60, 62, 141
Thirty Years' War, 32
Tinguely, Jean, 159
Tolstoy, Leo, 11, 79, 105
Toynbee, Philip, 125
Trakl, Georg, 8, 47, 86, 89–91, 94
Turgenev, Ivan, 22–4, 28–9, 78, 103, 115
Tzara, Tristan, 94–5

Vaneigem, Raoul, 93–4
Van Gogh, Vincent, 87
Vattimo, Gianni, 14, 64, 150, 152–60
Verlaine, Paul, 47
Vietnam War, 62

Virilio, Paul, 140
Vorticism, 92–3

Wagner, Richard, 8, 43, 58, 80, 83–5, 87
Wannsee Conference, 61
Weber, Alfred, 27, 60
Weber, Max, 4, 150
Wellek, René, 141
Werfel, Franz, 110
Wilhelm I, 23
Wilhelm II, 23
Wolosky, Shira, 135

Zeno, 43
Zola, Emile, 78–80